GOODNESS KNOWS

HOW!

BOOK 2

JAY LUMB

First published in February 2017
under the pen-name Jay Shaw.

Second edition published in August 2017
by jaylumbstories@gmail.com

ISBN: 978-1-9993480-1-4

Printed and bound in Great Britain by
CMP Digital Solutions

DEDICATION

To Mike, who would have loved Monrosso: the
Palladian house, the mountains, the lake, the wild life,
the trees and the animals, especially the wolves.

CONTENTS

MORE CONTENTS

THE STORY SO FAR

This book is a sequel to BOOK 1: GOODNESS KNOWS WHAT! Justin Chase has discovered the Marquis and his lady buried in the burned-out ruins of their palace in Rome. He has stumbled on a secret they are determined to conceal. Should they let him live - or might he tell the tale?

****

"What have you two done that makes you so keen not to be outed?" he ventured cautiously.

"It's not what we've done, it's what we ARE. You've just told us that you've seen that for yourself," said Lucrezia.

"So, WHAT ARE you?" he whispered hoarsely, "Aliens?"

They took deep breaths and hesitated, then gave that familiar Italian gesture of helplessness.

"Goodness knows!"

"We didn't arrive with labels round our necks," said Cesare.

"And we've never found anyone to ask," said Lucrezia. "But if Big Pharma ever get their hands on us they'll chain us up in cages in secret laboratories. They'll murder us in every way possible, trying to find out what makes us tick - so - you see - "

"What price immortality?" asked Cesare.

End of Book 1

He shifted his gaze to Justin's neck, to a spot just below the ear.

Justin flinched. Was that the spot where Cesare was about to land the edge of that powerful hand - the hand that didn't need a knife?

"You're right!" he gabbled desperately."Absolutely right! We've got to stop Big Pharma nabbing you. I'm with you all the way! Count me in! We've gotta stop them, or we'll be knee deep in Hitlers. There'll be Genghis Khan's coming up through the floorboards."

Cesare stared at him wide-eyed, then shrugged and turned to Lucrezia.

"Is he loco? Does he often rattle on like this?"

"We were talking about Japanese Knotweed last Sunday. He's not crazy: he's just a bit autistic."

"It's going to smother the world," said Justin. "There's no stopping it because it seems to be immortal. Immortality? We don't want it. It's a terrible idea."

"That so?" said Cesare. There was a hint of a grin breaking through and his eyes were twinkling. "But how many people would agree with you, do you think?"

"Lots wouldn't, would they? All the wrong people would bust a gut, give all they'd got, to become immortal. The world would be Hell. We've got to stop them. We've got to hide you. Anyway, last night you asked me to make you breakfast today. You don't want to miss that, do you?" The sweat was dripping off his chin as he watched the powerful hands.

"The full English," said Lucrezia. "He's good at that."

"The full English?" asked Cesare. "Nice lean bacon?"

"Of course! What else?" His heart was pounding like a steam hammer.

"Well, if you put it that way - Let's go! Lead me to it."

CHAPTER I

WHERE THE HELL'S THE BACON?

Justin held a shaking hand in front of the door mechanism. He bit his lip as the intimidating black doors of his penthouse flat swung open.

"Come in, come in, make yourselves at home!" he exclaimed in a loud, over-hearty voice. "And it's great to bring you back here looking so much better than yesterday. You were coughing yourselves to death. And just look at you now: fit as lops! What a relief! You really had me worried."

"We owe you a lot, don't we, Cesare?" Lucrezia said pointedly, looking her husband firmly in the eye.

"Mmm – and I could murder that English breakfast you promised us, especially the nice lean bacon." Cesare was rubbing his hands in anticipation, but then the gesture changed: the edge of his right hand sawed the left palm momentarily.

Justin winced. The body language seemed too obvious. Cesare was still itching to land that fatal Karate chop just below his ear. He put on the nearest thing to a laugh he could muster. Just murder the breakfast, he thought. You don't need to do the same to me.

"Well, breakfast coming soon. Just relax while I fix it."

Lucrezia followed him into the kitchen but he waved her away. "You two must have lots to talk about, so just leave me to it. You can set the table, then relax."

He let her pick out the cutlery and side plates, then shooed her out of the kitchen. "Go on, just relax."

When she settled on the sofa next to Cesare, he gingerly flicked on the intercom. It seemed wise to be forewarned if Cesare was still tempted to silence him for

8

good. Knowing Cesare's love of food, maybe he would blithely eat his breakfast, lick his lips, then coolly knock the head off his host. After all, his famous ancestor had a reputation for every kind of scheming skulduggery. Ancestors? How did immortality square with ancestors? Only two days ago he'd been eager to prove he'd had two descendants of the Borgias in his flat, and now - ! Crikey Moses!

Breakfast. Concentrate. What if he didn't have the right ingredients? Despite the softly whirring air conditioning, a hot flush of sweat broke out again. Bacon. Did he actually have any bacon? He looked frantically around his million Euro kitchen. His glittering Swarovski crystal central island was unlikely to save him if he couldn't rustle up the goods. Eggs? No problem. Baked beans? Several cans. Nice tomatoes. Fine. Where the Hell's the bacon? Must be some somewhere. Gotta have bacon.

The bakers on the corner would have some, wouldn't they? They'd need it for the pizza toppings. Should he slip out and get some? Might they think he was trying to do a runner, shop them to the police? If they took it badly he had not a cat in Hell's chance of outrunning Cesare, even with a head-start down the lift. That great ape-man would come leaping over the rooftops and drop on him from above.

Bacon! Thank Heavens for that! There was a pack lurking underneath the cheese. Saved by the bacon! Where's the frying pan? His hands were shaking so much that he dropped it. It hit the grey floor tiles with a resounding clang.

"Are you alright, Justin? Are you sure you don't want me to help?"

"No, no, I'm fine. Right as I'll ever be. No harm done."

Fine, my foot! he growled to himself. Would a normal red-blooded man be trembling like a terrified rabbit? Take deep breaths and relax – that's what the autism specialist always said. In, two, three; out, two, three, in - Look here, Mrs Trulove, how relaxed would you be with a pair of Immortals in your lounge debating whether to kill you, while you cook them a full English Breakfast?

"Justin, do I have time to ring our accountants before breakfast?" called Lucrezia. "May I use your phone?"

He flicked on the extension in the kitchen, and heard her on the living room phone explaining their predicament to their accountants.

They had been too badly affected by shock and smoke on Tuesday to talk to anyone, but now they were much better and being looked after by a friend nearby, but with all their credit cards and paper work turned to ashes they needed advice on proving their identity to their banks, since it was years since they had paid a personal visit.

It sounded exactly what any rational person would expect to hear – but it was mostly lies. It always was with Lucrezia, wasn't it? He'd discovered that days ago. Now that he knew why she told lies could he blame her?

Imagine the reaction if she told the truth! Burned to death in front of his eyes on Tuesday afternoon. Thursday morning, dug out of the rubble in a bad way. Died again last night, this time of lung damage, then woke up this morning as fit as lops! Well, we're just a couple of little old Immortals. Wow! Nice work if you can do it. What would you pay to pull off tricks like that?

They were absolutely right to be afraid of Big Pharma. And they had a lot to lose if they were rumbled. No wonder Cesare was tempted to silence him for good. Could he blame him?

Was the bacon too crozzled, the eggs undercooked? Did he prefer them over-easy? Or was he such a greedy-guts he'd happily scoff them any which way? It was a huge effort to hold the plates steady in his shaking hands as he set them on the dining table.

10

"Looks good!" enthused Cesare, piling in immediately with a big grin.

How different from yesterday when they were coughing up their bleeding lungs as they struggled to eat the invalid food he'd made them. They both seemed in such good health and spirits now that his fear began to ebb away. And Lucrezia did seem to be on his side – amazing, after the disgraceful way he'd treated her!

He watched their hands as they manoeuvred their food: perfect as waxworks. All the cuts and scratches caused by their struggles to get out of the rubble yesterday morning had gone without a trace. How did these miracles work?

"Do you use your special gifts to cure your patients or only what you've been taught at medical school?"

"Special gifts? Do we have any special gifts?" She looked puzzled.

Justin gasped and spluttered with amazement. "Do you have any special gifts? Like Hell you do! Surely rising from the dead perfectly cured is a very special gift in anybody's language."

"A gift? It's a curse!" growled Cesare. "You should try it!"

"Imagine being burned at the stake, knowing they'll do it again if they catch sight of you tomorrow, burn you again – and again." Her voice was rising hysterically.

Cesare grasped her hand across the table. "She's been torched seven times," he said quietly to Justin. "And it's not a very nice way to go."

"That's barbaric!" Justin's breakfast was trying to choke him. " Who did that? Why on earth?"

"Nasty twisted monsters. Some pretended it had some-thing to do with religion; some claimed red-heads were witches. Some just wanted to watch her suffer because she'd turned them down."

"I try not to be noticed, but men seem to target me."

"Is that why you gave that sexy white frock and those Louboutin shoes away? You know, the clothes you were wearing in Dubai."

They looked at each other and laughed.

"You mean the clothes they found on the roof of the Burj Khalifa entrance lobby. Wonder what forensics put in their report," grinned Cesare.

Justin stared at them open-mouthed. "You're telling me there was no safety net? You jumped off that balcony - "

"And hit the lobby roof going nearly 300 kilometres an hour. Splat! Great sport! Lucky we didn't hit head first and punch holes in the roof. That might have generated some headlines. Lucky it was so dark nobody spotted anything till our bodies had melted away. Come to think of it, that lobby must be armour-plated - -"

"You're crazy! How can you hope not to be noticed if you pull ridiculous stunts like that?" Justin exploded.

"You don't think we did it for fun, do you? You would have let them cripple you with gunshot and drag you off to be tortured to death, would you?"

"Mmm, well, if you put it like that - I honestly don't know what I would have done. Never been faced with that kind of choice – not in real life, anyway. I've put loads of crazy dare-devil things in my computer games, but they're just - daydreams, I suppose."

"You don't think about doing things like that for real?" There was a distinct smirk on Cesare's face.

"I don't have your physique. And I'm not immortal. I'm just your average, very mortal coward. My computer games are the only kind of rough stuff I can rise to." Well, he thought, most men are weedy drips like me, and we weedy drips tend to live a lot longer and stay in far better shape than the heroes.

"And how would I have fared if I'd tried to do the same as you these last two weeks? The Burj Khalifa trick would have been the end of me. I wouldn't have been around to die in the fire, or cough myself to death last night, would I? And what about Waziristan? Whoever saw off that terrorist chieftain didn't survive, did he? How many times have you died in the last couple of weeks?"

Should he be taking the bull by the horns like this? His heart was steam-hammering again. But the suspense was gnawing at his entrails. He needed to get a fix on what, if anything, his future had in store.

Cesare began counting. "One terrorist chief, two suicide bombers, the Burj, the fire times two - six. What about you, Krish? Three with me – any more?"

She winced and took a gulp of tea. "Just one more. The fanatics in Fallujah."

"They killed you, those fanatics in Fallujah? When? Why?" Justin gasped in horror.

"Two weeks ago today, the day before you first talked to me by the fountain. I wasn't very coherent, I suppose. Sorry. Still a bit shaken. Couldn't think straight. Just said the first things that came into my head."

"Good grief! And I treated you so badly. I didn't know. If only you'd told me," wailed Justin.

"What could you have done if I had told you? And you treated me exactly right. You made life feel safe and normal. It was a perfect cure: no responsibilities, just nice ordinary things to do, and in this lovely flat. It's too white and perfect for bad things to happen here. I'm so grateful to him, Cesare. We should think of some way to thank him."

"Well, yes, I suppose we both owe you a lot. You do understand, though, why we were a bit - uneasy - when you caught us waking up – resurrecting - this morning? It's hard to keep a low profile, and Big Pharma -"

"You'd be worth billions to Big Pharma. But you've no need to worry about me. I've already got millions more than I know what to do with, and I'm dead against helping Big Pharma create more Immortals."

Suddenly a thought struck him. "Are there any more people like you around, or are you unique?"

"There were thirty of us once," said Lucrezia. "About twenty men and ten women. But of course there may be others we've never come across."

"Haven't seen most of our people for thousands of years. Maybe they've found a way out, or they're very good at lying low," mused Cesare.

"Would you recognise them, after all that time?"

"Don't know," said Cesare. "How many billions of people are there in the world these days? Needles in haystacks."

"But there must be some way of recognising Immortals. They must look different in some way, surely."

"Do we look different?" asked Lucrezia.

"A few days ago I'd have said no, but now I've seen more of you I think you do. Maybe it's only when you've just been reborn, like now; haven't had time to get any knocks and scratches. You look too perfect, almost like waxworks."

"See what you mean." Cesare stared hard at Lucrezia. "Spose we ought to keep our distance from people at first, till we get a bit – worn. Thanks for the insight. In the old days we got dirty in no time."

"Glad to be of use," Justin smiled cautiously. "I told you yesterday you can stay here as long as you like. It's no inconvenience to me, and I'm enjoying your company. It's not every day one gets a chance to cook breakfast for a couple of Immortals. I'll be leaving for Rio a month from now on business so you can have the flat to yourselves for a few weeks while I'm away."

"That's incredibly generous of you, but I don't think you realise what you might be risking," laughed Lucrezia. "My darling husband is barely house-trained, so I shudder to think what he might get up to in this lovely white flat."

"Hey, what's that supposed to mean," laughed Cesare. "Are you calling me a peasant? What do you want me to do, dress for dinner every night?"

"Mmm, yes. You look very sexy in a dinner jacket."

"Black or white?"

"Have to be white in this weather, but you could wear a black shirt. You could have them sent from Monrosso."

"Or save them the trouble and go to Monrosso instead. How soon can we sort things out here, Oh Wise One?"

"Goodness knows," she said, wrinkling her forehead. "Medici's or Vetriano's could contact the insurers and the Heritage Department for us. They may take years to decide what's to be done with the ruin. We can't hang around waiting. We've spent so little time here for years we shouldn't miss it, and I'd grown to loathe the house."

"It's been a burden for years," growled Cesare. "The Authorities throw spanners in the works of every change we try to make, and who wants to run a museum?"

"Well, you've got a home here now whenever it suits you," said Justin.

"You're very generous, and I promise I'll keep the peasant on a short chain while we're here," she grinned, rumpling her husband's black wavy mop.

"Will you now?" he grinned. "Well, Justin, if we're staying here we'll have to learn to treat the place properly. It's very smart, so it would be a shame to trash it. Krish is right. I'm not up to speed in a fine modern home. Give me some firewood and I'll butcher and roast you an ox, or anything else I can run down or trap, but I doubt you'd appreciate that. It's about time I learned how to live in a normal home in the Twenty First Century.

Show me what to do. If you leave us here alone we'll have to do it, wont we? Better learn the way you like it."

Thank God for that! thought Justin. He had no choice but to offer them all he had if he wanted to save his own skin, but the thought of a big rough warrior mangling up his gleaming white penthouse was painful.

"So, where have you been living if you haven't been living in a house," he asked, as he demonstrated the correct way to stack the dishwasher.

"We've lived in castles, palaces, country estates for the last eight thousand years. It's not done to go barging into the kitchens. It scares the servants and offends your butler and your steward. And from long before that I've lived half the time in a war tent, or any rough shelter. The camp cooks feed you royally, but on the run you eat whatever you can get. If making smoke is too dangerous you eat it raw. Use your teeth to get the fur off if you don't have a knife. The last five years or so we've been living in tents or ruined buildings, glad to eat whatever the helpers managed to come up with. There are always casualties waiting, day and night. You can't let field hospital surgeons waste their time on cooking. So, we've missed out, haven't we?"

"You're joking!" gasped Justin. "You're walking history books. How can you cope with so many memories in your heads, so many skills you've learned? It's dizzying just trying to imagine it."

"Well, that's life," sighed Cesare. "We'd cop out if we could but it seems that we can't, so, what now, Master Chef?"

"Well, you watched me tidy up, ready for the next meal, so that's it for now. We could find something for lunch in the freezer, and get it out to thaw. Lucrezia, would you like to come and help choose lunch?" he called.

"I've just talked to Vetriano's and Medici's. They're

16

going to fix things for us, and I've an appointment to see the bank manager at eleven. I don't think we need hang around for long. Why don't we all go to Monrosso? Justin could trek in our mountains instead of the Himalayas."

"Yes, I insist," said Cesare. "You've made us very welcome in your lovely home. The least we can do is reciprocate. No quibbles: I insist."

Yes, I did see that coming, thought Justin. In his shoes I'd have to do the same – at the very least. Oh, dear!

CHAPTER 2

A BIT OF LUCK?

"Well, that's a bit of luck," said Cesare, "Gino has a helo he wants delivered to Lugano."

"Really? When?" asked Lucrezia.

"Any time soon. How soon can we get away, Oh Wise One?"

"Well, I can't think of anything else we can do here now, can you? I asked the bank to send our new cards to Monrosso. I told Medici's and Vetriano's that we planned to go there very soon, so they'll know where to find us if anything else crops up. Will it take you long to get ready, Justin? We insist that you come with us. You can go walking in our mountains as much as you like."

"And get eaten by your wolves?" Justin laughed.

"Never fear: we'll be there to make them cough you up again," she teased. "What kind of chopper is it? Is there room for all three of us?"

"Old R44. Pretty small. Room for two and two midgets sort of thing, and he's some big spare parts he wants to send along with it."

"Well, then, it's going to be a bit of a squeeze. Do you have people to see before we leave, Justin?"

"Well, I didn't expect to be here at the moment so I've nothing in my diary. Ready when you are. Where do we find the plane?"

"Urbe. Couldn't be better, could it?"

"Where's Urbe? Never heard of it."

"Only about seven kilometres away. Perfect."

"How long is the flight?" asked Justin. His old dread of flying seemed to be reasserting itself.

"Oh, ages," moaned Lucrezia. "Tiny choppers can only manage a little more than two hundred kilometres an hour. The Red Arrow trains are faster."

"Three or four hours, I guess," said Cesare. "We'll have to refuel on route."

"Well, we might be able to get a mug of tea at the pit stop. That would cheer you up, Justin," smiled Lucrezia.

"No room for a suitcase," warned Cesare. "No luggage locker, and the spare seat is going to be piled up with spare parts. Have to be just a small bag."

"Lucky we've no luggage at all, then," said Lucrezia. "Everything we owned here is burned to ashes. Are you happy to fly, Justin? Or should we go by train?"

"I can't. I promised Gino I'd deliver the thing, so - "

"So you can't let a good friend down. I know. Can I fly it, then, at least part of the way? It's ages since I flew a chopper. My licence will soon go out of date if I don't - "

"Justin can take a turn as well. We'll all fly it. Help to pass the time."

"Right, then," said Justin, swallowing hard. "What time are we off?"

"Straight after lunch?" suggested Lucrezia.

"Mmm," said the other two.

"Now, think hard," instructed Lucrezia. "When we get to Monrosso will we be thinking, 'Oh dear! We ought to have done this or that?' Whatever it is, let's try to think of it now and get it done."

"I ought to go down to the basement and see if - "

"Yes, you certainly should - "

"Oh, the McLaren!" exclaimed Justin. "What state will that be in?"

"The buyer took that away last Tuesday morning, thank Goodness. I gave it a good thrashing for him very early in the morning. The Panthers were half asleep, but we soon got them moving."

"You didn't show the buyer the way home through the mines, did you?" asked Lucrezia. "How did you escape?"

"I pocketed the cheque before we set off. When the car had him hooked I jumped out and left him to it."

"You left him to be arrested?" gasped Justin.

"They never press charges against a Saudi prince. He was happy as Larry I'd shown him what the car can do. He'll be bragging about giving the Panthers a run around. We teased out the Huracan - and a Gallardo."

Lucrezia pulled a face at him, and they both laughed.

"You have to pack, Justin," she said.

"And one of us should work out the route and locate a pit stop. Lost all my air maps in the fire. Why don't you do that, Justin? You're the computer freak. Krish can tell you what to look for. I'll deal with the basement." He stuck his feet into the big flip-flops Justin had given him and set off immediately.

Justin opened his wardrobes and stood scratching his head. What were you expected to wear at a Marquis's country seat? Did they really dress for dinner? Could he stuff a dinner jacket amongst the spare parts in a tiny little chopper? Would a clip-on bow tie be just too naff? He'd definitely need his hiking boots, and some decent shoes. The shoes alone would almost fill a small bag,

"Can I help?" asked Lucrezia. "It's hard to know what to pack to visit friends, isn't it? You never know what they're planning. You might be tramping around in the rain, with the dogs pawing at you, or they might expect you to posh up for endless pretentious parties."

"Yes, please. Haven't a clue where to start," he sighed.

"Shoes. Something comfortable to hike in, and slip-ons to potter about the house and grounds. For anything else you can just raid Chez's wardrobes. He's bigger than you so everything should be a comfortable fit."

"Would he mind?"

"Course not. He has lots of evening wear if we're invited out. Most of it's aeons old, but it still looks very glamorous, turns heads."

"Hmm," breathed Justin. "With his looks anything is bound to look glam, isn't it?"

"Well, you're not too bad-looking yourself, are you?"

"Haven't a chance against him, have I?"

"Nobody has, so don't beat yourself up about that," she laughed. "Just pack some undies and an extra pair of jeans and a couple of T shirts. And socks for the hiking shoes. The guest suites are like yours, everything's provided, so you don't need any toiletries."

"Just a razor," said Justin.

"Grow a beard," she laughed.

Eventually they sat down with Justin's satnav.

"Let's look at the whole route first," said Lucrezia. "Get our sense of direction. Put in Rome to Lugano."

"It's nearly the same route we took last week in the Black Arrow, at least as far as Genoa," said Justin, "but we were flying very high, forty-two thousand feet. It would be great to see the coast from much lower down, don't you think? How high will we be flying?"

"Only about two or three thousand feet," said Lucrezia. "So, we could set the compass to 290 to Civitavecchia, then hug the coast almost up to Genoa, then it's over land from there, heading 360, due north, for Milan and Lugano. That route will add some extra kilometres so we'll need to stop to refuel. Milan's three quarters of the way, so we need a pit stop south west of there."

"Milan Malpensa is around there," suggested Justin. "There we are."

"Sorry, Justin, no go," said Lucrezia. "It's a huge busy airport. We'd have trouble getting a landing slot at the last minute and buying a small amount of fuel might be very complicated."

"Milan Linate?" said Justin.

"Even worse," wailed Lucrezia. "Everybody complains it's a shambles and it's nearly as big as Malpensa. We could be stuck there all night – if we managed to get permission to land. What we want is a quiet little landing strip with a fuel supply. Could you persuade Google to cotton on to that."

After a lot of sparring with Google, Justin at last found Voghera.

"Looks good. Flying club, open till seven tonight. Fuel? Yes, Jet and Avgas. Just the ticket." said Lucrezia. " If Chez is happy with it, I'll ring for permission to land. Now, you'd better finish packing while I make lunch."

Cesare was frowning when Justin let him in. He stood meekly just inside the hallway dripping dirty water.

"The looters have been making hay already."

"Was there much worth taking?" asked Justin.

"Well, they seemed to think so. Thought I'd have to move the rubble out of the basement staircase, but they'd done it for me. As far as I can tell, they've found precious little that matters to us down below, but they've hacked out the copper pipes and the water was running everywhere. Managed to find the main stop cock and shut it off. They've started sorting through the great heap of rubble we moved last night, flung it all over the place, covered up the bed. Sorry about the mess I'm in. There was burnt ash and water everywhere."

"Well, you've time for a quick shower before lunch. I'll find you some clean togs." He cleaned up the flip-flops: he had nothing else that Chez could get his feet into.

When Cesare had dressed he steered him back to the

entrance. "Let me set the door control so you can let yourself in and out. Hold your palm just here."

Cesare held up his palm and stared at him quizzically.

"Right," said Justin, after pressing a few buttons, "take your hand away; now put it back again. Oh, doesn't seem to have worked. Let's try again."

After three unsuccessful attempts, Cesare sighed.

"Can Krish make it work?" he asked.

"Lucrezia, can you come here a minute?" Justin called. "Hold your hand up to the door mechanism."

She did, and to his surprise, nothing happened. He reset the mechanism but still it wouldn't work for her.

"However do you get in and out?"

"The first time it wouldn't work I saw you press the override button, the one hidden behind here. I use that."

"Well, I suppose you'll both have to use that."

Lucrezia took Cesare outside to show him the other hidden button.

"Funny. Anna, my help, has no trouble with it. It must be faulty. I'd better get the technician to come look at it when I get back," said Justin.

"I suspect he'll find no fault with it," said Cesare, sadly. "It's just us. We're freaks. Can't travel on a plane without a lot of trouble at Security. Finger prints and iris scans are out. Heaven help us if everyone starts checking DNA like Kuwait!"

"Crikey Moses!" exclaimed Justin. "You've got a serious problem!"

"Yep," said Cesare. "Another of our special gifts."

"Lunch is on the table," called Lucrezia. "Did I hear you say the rubble was back on top of the bed? Had we better shift it before we go?"

"Happy little optimist, aren't you," grinned Cesare,

tucking into his lunch. "The looters have only taken the top off the pile, so I bet they'll be back to finish the job in stages. Do you want me to hang around indefinitely, endlessly shifting rubble, just in case we kick the bucket again in the near future?"

"Well, I admit the last two weeks have been abnormal. We usually manage to stay alive at least a few decades at a time. You could ring Medici's, ask them to persuade the Authorities to let us get the rubble taken away. They might take more notice of you. I can't think of a good reason. They'd just write me off as a hysterical female, say, 'Yes, yes, My Lady,' then do nothing."

"The insurers will insist on checking the ruins for arson. If we'd managed to get the rubble removed beforehand they might try to charge us with burning the place down ourselves and removing the evidence. We have to stop thinking about it. Get on with life. I will ring Medici's and tell them about the looters, so they know we haven't been tampering with the evidence. You won't talk to anyone about digging us out, Justin, will you? And moving all the rubble last night."

"Of course not. I'm just as keen as you are not to give the game away. It's the Great Four Poster Bed you're worrying about, is it?" asked Justin. "Why do you always reincarnate yourselves in that?"

"Goodness knows!"

"Every time? Didn't you say something about dustbins in Waziristan?"

"Peshawar, yes. I woke up three times behind those blasted dustbins at the American Consulate the other week. And no. I've not the faintest idea why."

"Cesare, you've got to think hard about it. Something must have made that happen," urged Lucrezia.

"The Borgia House was only five hundred years old. If you've been around since the Stone Age, you must have been reborn somewhere else for thousands of years before it existed," argued Justin. "You surely don't want to be tied to the ruin. You need somewhere new, with privacy and clothes in the wardrobe like you had in the B House. You're very welcome to use this flat as your new Rome base, but how can you arrange that? Was there anything special about the places where you were reborn? What kind of magic do you have to use to make it happen?"

"Magic? Big con," laughed Cesare. "It's very decent of you to offer us this great place to home back to. As to how we can arrange that, your guess is as good as mine. It's hard to remember much detail about resurrecting in the old days. I've pictures of so many places whirling around in my head, mostly battle fields, struggling out from under piles of corpses. Beds? Yes, but which was where will take some working out."

"Yes, you're right, Justin," said Lucrezia. "We've got too complacent over the last five hundred years. Waking up at the B House was so convenient, just routine. Pick another frock out of the wardrobe, then sail out to carry on living as if nothing special had happened. We've got to find a way to resurrect somewhere sensible, or who knows where we might end up."

"Why don't we let that soak? We've a boring flight ahead so our subconscious minds might amuse themselves by working it out. Now, what needs doing before we go? If you can give me an ETD I can ring Gino."

CHAPTER 3

POOR OLD CHOPPER

"This place gets better every time we come," said Lucrezia, as Cesare paid off the taxi. 'AEROSTAZIONE ROMA URBE HELO HUB'. I like this."

Justin surveyed the smart new building, and the long runway behind it.

"I've never heard of it before. Amazing they found the space for a new airport so near the city."

"No, it's the oldest airport in Rome, built by Mussolini. Rome Flying Club had to campaign for years to keep it open when they moved the passenger planes to Ciampino. Nearly got turned into an industrial estate. With this new helo hub its future looks much rosier."

"Good place to keep your executive jet," said Lucrezia. Private planes dotted the grass like daisies. "Perfect site. Easy to spot from above with the road and railway down that side and the Tiber snaking around the other three. Nice approach and take-off." She led the way inside.

"Where can I find Gino Volpe?" asked Cesare.

"I've a note about him somewhere," said the receptionist, rifling amongst the papers on her desk. "Ah, here it is. Mr Volpe sends profuse apologies. One of his pilots phoned in sick, so he's had to fly the customers himself. Back about six or seven this evening. Luca Verde will see to everything, Sir, the Chief Technician. He's over in the workshop, the big hanger over there. I'm on my own here -"

"We can find our way. Don't worry," said Lucrezia. She led the way out onto the apron and across to the open hanger, then ploughed confidently in amongst the dozen or so small planes and helos on trolleys.

26

This was the first time Justin had ever seen an aviation workshop. The planes, a bewildering tangle of wires and pipework, were barely recognisable with their metal skins stripped off, so he tucked himself close behind her, wary of doing himself or the planes any damage.

"Luca Verde?" Cesare shouted over the din. Arms pointed, men shouted, and eventually a man in a boiler suit came towards them, wiping his hands on a rag.

"You must be the Marquis. Glad to meet you, Sir. Gino was really pissed off about missing you, but he didn't have any choice. No one else available to fly the plane. He told me you're going to fly that old R44 up to Lugano for him. It's over here, on the grass."

Pretty little thing! thought Justin, as they walked across the grass towards the chopper. Looks more like a Christmas tree decoration than a vehicle — far too tiny for three hulking adults, surely. The small glass fish-bowl masquerading as a passenger cabin looked like something that might turn into a pumpkin on the stroke of midnight. Behind it stretched a very long, very thin tail painted the brightest shade of yellow.

Verde pointed to the back seat where a large lumpy shape, wrapped up in an old towel secured by shock cords, had the seat belts fastened around it. Cesare nodded and gave it a few exploratory shoves in each direction, then tightened the seat belts a little.

"What's that?" Justin asked Lucrezia.

"Spare parts to do up the chopper, I gather," she replied. "Can't afford to have them flying around loose,"

"Where's the maintenance log book? demanded Cesare.

Luca Verde extracted a grubby book from the pocket behind the seat. Cesare studied it carefully.

"It looks rather battered," said Lucrezia, pointing to the dirty scuffs and tears on the spartan seats.

"Jeez!" exclaimed Cesare, "Gino did well to sell this old kite in this condition. It's done so many hours EASA'll be on the point of revoking its licence by the time we get it to Lugano. Does the buyer know what he's in for? The overhaul's going to cost him an arm and a leg."

"Yes, it's just what he wanted. Asked Gino to look out for something just like this. He was a mechanic at Lugano airport till a few weeks ago, but he's had to take on the family farm. He wants to do this up to keep his hand in."

"He should come and fetch it himself; check it out."

"Can't do that," said Verde. "Dad's in hospital, mum's hysterical, cows need milking twice a day."

"Fair enough. So, you think this thing might actually fly?"

"Of course it will fly. Gino flew it up here yesterday himself. Got it in an auction south of Naples. He's had us working on it all morning; said it had to be in good nick for you. Bearings whining a bit, of course -"

"Never met an old R44 where they didn't," shrugged Cesare. He leapt up onto the door frame and reached for the drooping blade, running his hand along the edges as far as he could reach.

"I'll get you a ladder," said Verde.

Cesare walked around the tiny bauble, noting the dents and scuff marks on the paintwork and the scratches on the glass. He opened up hatches on each side of the body and poked about inside. When the ladder arrived he felt the whole length of the tail rotor blades and then the main rotors.

"I'll waggle the controls for you," said Lucrezia. She climbed in and began fiddling with something.

"I can't see what she's doing," said Justin.

"Watch the rotor blades," said Cesare, from the top of

the ladder. "They swivel, see? Face the air at different angles, make it behave in different ways. Swash plate looks okay." He pointed to the complicated mechanism below the main rotor. "Fuel?" he asked, as he handed the ladder back.

"Full right to the top. Done the water-in-the-fuel check just ten minutes ago."

"Avgas?"

"Yes; doesn't get on with Jet."

"Shall I lift her off, so you can see if she looks okay – and sounds okay?" called Lucrezia. "Clear rotors!" She waited till they had backed away. Cesare grabbed Justin and dragged him well away from the long, skinny tail.

It felt like a roaring force ten gale. Justin covered his ears and leaned forward to avoid being blown over. He watched as the tiny yellow chopper lifted off the ground, wagged its tail from side to side, then spun like a merry-go-round. Then it flew forwards, backwards and round in a circle. Amazing machine! At last she eased it down to exactly where it had been and switched off the engine.

"Great pilot, isn't she? Okay?" he called.

"Mmm. 'Spose so. I'm game if you are. What about you, Justin? You and I could go by train if -"

"If you think it's okay, I'm up for it," he smiled weakly. It would have been great to sit back on a nice safe quiet train. Man up, you great wuss! he yelled silently. Cowards are masochists. Get in this half-dead chopper. Face what comes. It wouldn't seem very friendly to leave Cesare to fly there on his own.

"Who's sitting where?" asked Lucrezia, while Cesare walked back to the hanger with Luca Verde. "We'd better put Chez in front of the spare parts as he's the heaviest, and you and I on this other side, to balance the plane. Do you want a turn at flying her?"

"Would you think me a complete wimp if I sat in the

back? A dodgy plane with an ignoramus at the controls is a bit daft, don't you think?"

"Are you sure you don't want the two of us to go by train?"

"Would you rather do that?"

"Not really. Chez likes the idea of flying up the coast, sight-seeing for a change. It should look wonderful from two thousand feet up, quite different from your trip in the Hawker Hunter."

She climbed back into the front seat. "Did you bring your satnav phone? I hope Chez asks the mechanic for an air map. All ours were burned in the fire. We normally fly in a fairly straight line, and don't have to stop for fuel, but this little chopper only holds enough fuel for 550 kilometres. Once we're into the mountains beyond Milan we'll have to gain height and find our way around the peaks. We don't want to be worrying about running dry."

Cesare had been right to describe this as a 'two and two midgets' sort of plane. There appeared to be leg room in the front, as long as you were allowed to put your feet on the pedals, but there was no way to get comfortable in the back. His knees were not far from his chin. Four hours was going to feel a pretty long time.

"A flight map for you," said Cesare, tossing it over his shoulder as he settled into his seat. "You might find it interesting to try making sense of it. Right, final checks."

"Did them while I was testing her. Looks quite normal."

"Well then, beam us up, Scottie."

"Tell me what you're doing, how you fly these things," said Justin.

"Well, these pedals make the rear rotor swing the tail around. We've no brakes because we don't have any wheels. First we have to lift her off the ground with the collective, this lever down here, like a big old-fashioned handbrake. So, let's go. Clear rotors!" she shouted, had a good look around, then turned the ignition key.

"You haven't shut your door, Cesare," Justin shouted over the roar of the engine.

"No door to shut. They've lost it. Came from Naples. Probably the Camorra. Maybe they had a machine gun poking out, or threw out bags of drugs and ciggies."

"Should we be flying without a door?"

"Don't worry," she said. "Most people fly with a door off in this weather, otherwise we'd fry in this glass bubble. Hang onto the maps, though. Anything that blows out can hit the rear rotor and wreck the plane. Right, let's see if the Tower will let us go. "Good afternoon, Urbe Tower. Helicopter India Alpha Bravo Xray Kilo. Request clearance for take-off, heading to Lugano."

"Good afternoon, Helicopter India Alpha Bravo Xray Kilo. You're cleared for immediate take-off, main runway 34. Standard take-off heading north."

He watched her pull up the lever at the left side of her seat. The craft rose so smoothly that he didn't realise it was airborne until it began to move forwards.

"This handlebar, the cyclic, does similar things to the joystick in a plane, swings the nose up and down, sends us forwards, backwards and round in a circle, So, cyclic slightly forward, collective up a little more to get more height and speed. Come on, more speed. What's the matter? Your throttle switch, Chez. Is it uncoupled?"

He fiddled about with a button on the collective on his side. "Give me control. No, it's not working, dammit."

"Are we in trouble?" asked Justin.

"Not really. Faulty switch. It happens. The throttle should be linked to the collective so when you pull the lever up it gives you more power automatically. Twisting the throttle's just an extra chore. Probably we'll find lots of other little faults at her age."

31

The little battered chopper was now crossing the river.

"You have control," said Cesare. "What's your course?"

"290 as far as the sea, then follow the coast more or less to Genoa, then due north past Milan to Lugano."

"What about fuel?"

"We thought Voguera, south of Milan. No commercial traffic, just a flying club and a flight school. Avgas and Jet. Open till 7pm."

"Great! Easy to spot from the air?"

"Mmm. Not so sure about that," said Justin, anxiously. "We had trouble spotting it on Google Earth."

"Well, check it out on the air map. Start now, while the scenery isn't thrilling."

Justin opened up the grubby map, stiff and heavy with its plastic coating back and front. It was hard to make out the towns and roads with so many unfamiliar long black lines and hieroglyphics printed on it.

"These pink circles, are they beacons or -?"

"Small airstrips," said Cesare, "The smaller the circle the smaller the airport. And then there's the little black circles with no pink frill around them. Could be micro-lights, flying carpets, witches on broomsticks, anything."

"There seem to be hundreds of them."

"And at this low height we need to keep well away from all of them. I presume this course will keep us well away from Fiumicino?"

"We hit the coast just north of Civitavecchia, miles away from Fiumicino. Then we can just chug along, following the coastline more or less, up to Genoa."

"Find Rome Fiumicino, Justin. See the size of the area that tower controls, a huge black oblong with a wide purple fringe," Cesare instructed. "We'd be in trouble if we flew across that less than twenty thousand feet up."

At first Justin peered down at the ochre farmsteads,

the parched yellow grass, the dark green forests and irrigated crops, but there was so much of it all, and it soon became a bore. It was quite a thrill to see the sea, shimmering in the hot sunshine. She flew low over the sea along the front of the first resort, and a few sunbathers waved as they passed. Buildings looked so different from above that he couldn't recognise towns he knew quite well. One after another, sea-side resorts slid smoothly beneath them, while he struggled to match them with names on his map. A few resorts had pink circles nearby, so he could practise trying to locate the airports – far from easy.

"There's a pink circle on the map, and three planes down there in a field, but no runway. How come?"

"Grass runway, probably,"said Cesare."Quite common."

"How on earth can anyone spot it if there are no planes on the ground? The first plane back must have a devil of a job finding where to land."

"You've never been truly lost till you're lost at Mach two," joked Cesare.

"Have you ever been lost up in a plane?"

"You bet! Not nice. Can't pull off the road and ask directions."

"What on earth did you do?"

"Depends where you are. Head for the sea, if poss, then at least you're less of a danger to everyone else. Fly along the coast till you recognise somewhere. Unless you're concussed you must have some idea which area you're in. As long as the compass is working, or the sun or the stars are visible, you can work out which direction you're going. If you can read your watch you should be able to work out how far you're likely to have travelled since you got lost or fell asleep. Nowadays planes have satnavs, of course."

"So you always sort yourself out in the end?"

They both laughed, but they didn't reply.

Oh cripes! thought Justin. Wonder if they survived. How many planes have they written off?

"We were going to try thinking up some way you could control where you woke up, you know, resurrected." he said, after a judicious pause.

"Yes" said Lucrezia. "You were going to try to work out why you woke up behind the dustbins in Peshawar."

"How the Hell do I know?" Cesare grumbled. "Maybe it's a message - put me out with the rubbish. Next time they'll cart me off to the dump."

"Oh shame!" she laughed. "Come on, it's important. It took you ages to get home last time, and next time they might lock you up for years. Think hard. I know thinking hurts your lovely thick head but you'll have to be brave. What were you thinking while you were dying?"

"What the f*** do you think I was thinking? Aoouch!!!" he laughed. He shut his eyes and grimaced. "Well," he said at last, "maybe you're right, O Wise One. I was thinking, 'You bastards! Think you're getting rid of me? You're in for a nasty shock. I'll be back. Make you sorry you were born.' Didn't want to go home, did I? Not till I'd sorted them."

"What about the second time?"

"Well, obviously didn't think I'd wiped out enough of the fiends, did I? Stayed on to do another and another."

"There's hundreds of terrorist fanatics still out there. What stopped you?"

"The Yanks got sick of finding me behind their bins every morning. Insisted on giving me a lift home. Then the house burned down, burned all my gear."

"Otherwise you might still be there, playing avenging angel? There's far too many of them, you know. Like trying to drink the sea dry."

"I know, I know. It's hopeless. Might even be making things worse, making martyrs out of psychopathic killers. Who knows. Well, I've tried curing them and killing their monsters for them - what more can I do?"

"Take charge of the world again? Make them behave like civilised humans?"

"All seven billion of them? It was bad enough when there were only a few million, wasn't it? And how do we bow out? Who takes over when we've had enough? Last time it was back to chaos in five minutes. They were killing each other again and smashing up everything we'd created for them."

"Mmm," she sighed. "A few years ago I thought they were growing more civilised, just needed a helping hand, but maybe it's a forlorn hope, We can't save Humanity. We did try. And I'm so weary of it all."

"Well then, we should think about ourselves, for a change, sort ourselves out. Maybe all we need to do is *want* to be somewhere. Right now all I want is home."

Perfectly ridiculous Alice in Wonderland solution, thought Justin. Anything's possible in Wonderland, even taking over the world.

"Wake up, Navigator!" Cesare's voice bellowed over the din. "Genoa ahead. We need you to find Voguera."

Crikey Moses!, thought Justin. It was all too easy to fall asleep, battered by the appalling din of the engine and the throbbing rotor blades. What if he couldn't find Voguera? Head back to the sea again, and then what? No way do I ever want to be a navigator. But flying alone, like a fighter pilot, you *are* the navigator as well! No, you can't just pull onto the hard shoulder, can you? Ask directions, study the map and the signboards. What the hell ought he to do?

"Plenty of time," she said comfortingly. "I'm heading due north now so it should be pretty well on the nose. It's about seventy kilometres, so more than a quarter of an hour ahead. You could try the satnav on your phone, get a picture of the airport again."

With shaky hands he powered up the phone to key in Voghera on Google maps. Nothing. 'No service provider,' it said. It had worked on the ground in Rome. This thing was supposed to work like a satnav in the air, but nobody seems to have told it. Planes had satnavs that worked, surely. Well, this didn't. Back to the paper map.

The weather ahead looked threatening. A huge black hammerhead was flowing towards them. Ducking under the cloud, she tilted the nose downwards until the ground looked frighteningly close. How odd! He'd developed a fear of heights in reverse. It had felt much safer with a lot more distance between him and the ground.

"We're flying under Visual Flight Rules, must be able to see the ground at all times," she said. "Oh, nasty."

It was one of those violent summer storms. The heavens opened, rattling hail on the windows. Could she see where she was going? She was checking the dials carefully now.

"Don't worry, I can fly on instruments alone, and it's not as if I have much choice. Lucky they are all working."

The storm ended as suddenly as it began, but it must have eaten into that precious quarter of an hour. Back to the navigating.

It took him a while to single out Voghera's pink circle from others quite close by. Maybe one of the others would do if he couldn't find the right one. Reassuring. Maybe they could put this thing down plop near a petrol station out in the wilds if they had to, unlike a fixed-wing plane. Borrow a can and fill it up.

"Can this thing fly on car fuel?" he asked.

"No, it's very choosy. Won't tolerate anything but Avgas. Anyway, you can't just put a chopper down in a public place unless you're police or ambulance. We'd be in dire trouble. Might lose our licences."

Justin swallowed hard. He studied the map minutely, trying to soak up every detail. Wooded countryside lay below them now. On the map it looked to stretch almost to Voghera, where farmland took over. Maybe he could spot the change in topography. It was a wide belt of farmland edged to the north by a big river snaking around the fringes of mountains. If he saw that river and those mountains they would have gone past Voghera.

Now, how to find the airstrip needle in this farmland haystack. Roads. There was a very straight road on the map, running past the airstrip, linking Voghera to Rivanazzano. He could remember seeing something dark green on Google Earth, snaking along the other side of the airstrip. A river? There was a meandering blue line on the map. And here it was, as if by magic, the runway slipping past his window. "Stop! We're there," he whooped.

"Well, thank goodness for that!" she exclaimed. "Voguera Rivanazzano Tower. Good afternoon. This is Helicopter India Alpha Bravo Xray Kilo; request landing instructions, please."

There was a long silence. She repeated the call. No reply. "I booked it before we set off. Seems there's nobody in the tower now. What shall we do, land anyway, apologise later?" By now she had circled right around the airstrip. There were only two planes on the ground, parked right up to the buildings. "There's nothing moving so we should be okay."

Cesare jumped out as soon as they landed and headed for the tower. Glad to stretch his legs, Justin followed.

"Who's landed that bloody helicopter? What does he think he's doing?" yelled an irate voice. A red-faced man with a nasty scowl erupted from the building.

"Good afternoon," said Cesare pleasantly. "Is it in an inconvenient place? Where would you like it?"

"At the bottom of the sea, for all I care. Who gave you permission to land?"

"We telephoned this lunch time. Did nobody record the booking? We just now radioed several times and got no reply. We just need to buy some Avgas, then we'll get out of your hair."

"You're not a club member, are you? Not seen you before. Can't sell gas to non members."

"According to your airport information you advertise Jet and Avgas for sale, and are open to the public," said Cesare, mildly. "Is there some problem today? Seems very quiet here."

"You can say that again," growled the man. "Whole bloody club's gone off on a rally. Muggins here's drawn the short straw. Nobody asked me to crew for them, so I'm stuck here, minding the shop. Good mind to resign from the club."

"Hard cheese," said Cesare evenly. "That would piss me off as well. What's happened to the controllers?"

"One's sick and the other's - God knows! Nobody's due back till late tonight so maybe he's having a kip."

"So, is there anybody here who can sell me some gas?"

"Yeah. I'll go find him," he muttered grudgingly.

"Tea?" asked Lucrezia, sauntering up hopefully.

The men laughed. "You try asking him, if you want to get hung up by your toe nails," said Cesare. "The café must be closed. Everyone's flown off for the day and left one poor chap behind. He's spitting blood. Have a walk around and stretch your legs while I see to the fuel."

By the time they climbed back into the plane they saw Cesare patting the old man on the back and laughing.

The old curmudgeon grinned as he waved them off.

"Selling snow to Eskimos again," she smiled. "So, now, all we have to do is set the compass to 353 and Lugano here we come. There's no wind to allow for."

Justin breathed a sigh of relief. Drama over. Just try to forget the cramps and relax. Have another little nap - -

Aouch! What's that? The plane was suddenly lurching, nose down, nose up, shaking him violently up and down.

"Chugging! Dunderblitshun!" roared Cesare. "Give me control."

"You're welcome," she shouted. "What on earth is causing this? We're nowhere near overloaded. It was fine before."

"Lord knows! Nobody's got to the bottom of this. They say more speed usually stops it."

He twisted the throttle. The engine roared louder but the chugging was relentless.

Something dropped hard on Justin's knee. He fended it off and it slid down the side of the seat. Another heavy piece of metal got him on the shoulder. Auch! It had spikes. The parcel of spare parts was disintegrating, throwing off its restraints. A small one hit the ceiling and dropped onto Lucrezia. The revenge of the machines, thought Justin.

"We need more height!" shouted Lucrezia.

"I know. Something wrong with my collective. Won't come up. Is something jamming yours?"

"Yes, I can't get hold of mine at all. There's a lump of metal blocking it. It seems to be wedged under the seat."

"Well, get it out before we hit the ground."

"I'm trying to."

Justin tried to help, but couldn't reach it. He unfastened his seat belt and promptly hit the roof. Lucrezia squealed as he grabbed her hair along with her seat back and tried to get his other hand down the narrow gap between her seat and the door. He could see the lump of machinery wedged there but in the few brief moments before he hit the roof again there was little time to work out how to extricate it.

The chaos was dying down, the chugging decreasing, but all was far from well. The engine had caught a cold, coughing and spluttering.

"What now?" asked Lucrezia. "Are you sure they sold you Avgas? The engine doesn't seem to like it much."

"Could be water in the fuel. Everything was wet when we filled her up. Should have moved that rubble," he muttered.

"Hope they've emptied your dustbins," said Lucrezia.

"What about me?" squawked Justin.

"It's alright for you, Sweetheart. You'd just have a nice long sleep, but don't worry," she said soothingly, "we'll soon sort this chopper out."

"The map," shouted Justin. "Grab it, Cesare. It's flying out of the door!"

Cesare made a lunge at the map but missed it. Off it flew, while the chopper responded to his abrupt shift of weight with a new motion, a dip from side to side, alternating with fore and back.

He'd be sick if this carried on much longer. It didn't.

There was a loud bang. "What now?" gasped Justin.

The answer was not long in coming. The tiny craft began to spin, first slowly, then more quickly. With its tail rotor smashed by the heavy plastic map there was nothing to counteract the main rotor's spin. And it was eerily quiet. The fuel must have choked the engine. There was only the swish of the free-wheeling main rotor, throbbing more and more slowly, like a dying heartbeat.

"The river, thank Goodness!" gasped Lucrezia. "Dump her in the river."

"Do my best," said Cesare, through gritted teeth. "I'll get home as fast as poss to dig you out."

"No!" yelled Justin. "Monrosso's where you want to be. You've got to shout, 'Monrosso, here we come!'"

The doomed craft performed a spectacular swansong, spinning and tumbling as it lurched haphazardly towards the broad slow river. Justin was flung at the ceiling, then at the walls, like the clothes in a spin drier, then up over their heads and down into their laps. They wrapped their arms around him to try to hold him down. Bloodstains were spreading across his shirt and his shoulder was throbbing and smarting. What did it matter? He'd surely be dead in seconds.

"Monrosso!" he yelled like a maniac. "I wish I was going to Monrosso!"

"Monrosso! Monrosso! Here we come!" yelled the other two.

"Monros - - - -

The blow rang his body like a cathedral bell. The shock made him gasp, draw in searing fumes that scorched his lungs, fumes with the terrible reek of blazing fuel. He held his breath - -

41

CHAPTER 4

WOLVES

He needed air: he had to breathe. There was no pain this time. He tried another breath. The air was fresh and cool, infused with the faint scent of flowers. He dared to take a deeper breath, and the effect was thrilling. The air surged through his body, a tidal wave of exultation.

He had died and gone to Heaven.

But why? He had not made the slightest effort to ingratiate himself with any deity. Had he been granted a place in Paradise for free? He flexed his muscles to see if he still possessed a body. He had never felt so fit, so comfortable in his skin in all his life.

Sounds came floating in: first the hum of bees, then a new delightful sound – a bird, singing a song he'd never heard before, hardly a song at all, more like the sound of kisses. In Paradise the birds blew kisses.

He opened his eyes. There on a tree branch overhead sat a plain black birdy kind of bird, with a dark grey hood. On Earth the crows cawed like cackling old witches. Here they blew kisses.

What other delights awaited him in Paradise? The fresh green leaves on the tree above him looked the most perfect shades of green. The blue sky behind the flickering leaves was a perfect shade of blue, and the puffy fair-weather clouds looked straight from the cloud laundry, bleached a perfect, perfect white. Please don't tell me it's all a mistake; don't order me back to Earth, he moaned to himself silently.

Something odd was happening to his right foot. Hot damp air was pulsing onto it, and now something scratchy, like a Brillo pad, was working on it. Some

42

instinct told him he should investigate this very quickly. He lifted his head - and looked into a pair of slanting yellow eyes. A rumbling growl broke into an open snarl as he drew his foot away, and an evil-looking set of spiky teeth avidly followed the foot.

They had wolves in Paradise.

Suddenly something erupted from the ground beside him. Cesare stood, arms akimbo, glaring at the wolf.

"How dare you threaten my friend, you miserable beast?" he roared.

The wolf retreated a few meters in surprise, then advanced a little, snarling ferociously at Chez.

"Place your bets," said Lucrezia, sitting up and hugging her knees. "Man versus Wolf. Unarmed man versus fully armed wolf."

"Armed wolf?" queried Justin.

"Look at those great big teeth and claws. Big sharp daggers on every toe. What more could he need?"

"Well, Cesare's immortal so he's bound to win," said Justin dismissively.

"All that means is he'll wake up tomorrow, whatever happens today. He's as vulnerable as any normal human today."

"So, what happens if the wolf wins?"

"The pack will have the three of us for breakfast – unless you can take them on. We're outnumbered, six to three." She nodded towards the edge of the glade where five more wolves were emerging from the trees. Reaching out surreptitiously, she picked up a sizeable stone.

Cesare turned as if she had called him and caught the stone. He waved at the wolf pack, then swung his arm backwards and forwards until the wolves were all following his arm like fans at a tennis match. At last he flung the stone with amazing force away into the distance. Off went the pack like dogs chasing a ball.

The lone wolf bounded forward as if to join the pack, then charged at Cesare and hurled itself straight at his bare undefended back.

"Wooh!" breathed Lucrezia, "never turn your back on a wolf." She looked around the clearing, and reached for a sturdy fallen branch. "Don't move, Justin, but look for a useful weapon. He may need reinforcements."

Cesare must have sensed the wolf before he felt it. He was already twisting his body so that the heavy animal swung around and landed on its back on the ground with Cesare on top. There was a frantic struggle as the creature, snarling and yelping, snapped wildly at him with it's sharp spiky teeth and slashed at him with its claws.

At last Cesare managed to get his hands around its neck and squeezed. It wriggled frantically, fighting to breathe. He slackened his grip and let it squirm around onto its front, lying astride it, out of reach of its snapping jaws. The beast began to howl and whine and whimper.

"What's he doing?" asked Justin. "It doesn't sound very happy about it."

"Threatening to crack its ribs with his legs. It must be terrified."

At last he relaxed and began to talk to the defeated creature. "Behave yourself, you nasty beast," he snarled. "We lead, you follow. We feed you. You guard us, never harm us. Understood?"

He stroked the wolf's head, but it growled and tossed its head. "Don't you growl at me," he roared, feigning a blow to its head. He stroked its head again. It cringed silently. "That's better. Good dog." He stroked it again, then held his hand in front of its mouth. The wolf licked his hand. "Good dog! Now, come meet my folks."

The wolf slunk dejectedly behind him, ears down and tail drooping. He paused now and then to stroke it. Patiently it allowed Justin and Lucrezia to stroke it and licked their hands.

"Poor thing!" crooned Lucrezia, giving it a hug.

"What about me?" grumbled Cesare. "I save you from being a wolf's breakfast and the wolf gets all the sympathy."

"Oh, shame!" laughed Lucrezia. "You were fabulous. It was thrilling to watch you, and just look at all these dreadful battle scars. We should parade you through the streets, have a great procession in your honour."

"And offer me the juiciest virgins. Next time Justin can have a go, while I sit and watch. And talking of breakfast, let's go home."

"Dressed like this?" asked Justin. Enthralled by the fight, he'd not really registered their lack of clothing before. What on earth was he doing out in the middle of nowhere with not a stitch on? Yes, they were in the same naked state, but he'd seen them like that before – twice. It seemed almost normal for them. Maybe this was a sort of Garden of Eden -

"Let's look in the Refuge. We used to leave raincoats in there in the old days, in case the weather turned bad while we were up here."

The Refuge was a picture book stone cottage. Beside the huge open fireplace, scented logs were piled up high. Around the walls were built-in benches, just wide enough to sleep on, heaped up with cushions and travel rugs. Two oil lamps swung over a rustic wooden table. Perfect place for a snug little party, thought Justin. In winter with a roaring fire.

In the entrance hall there were rows of sturdy coat hooks, and many were in use. There was a wide choice of summer raincoats, but they proved a disappointment.

"You'd look great in this, Justin," teased Lucrezia, holding a transparent pink raincoat against him. "I defy you to start a new fashion in this."

"You will visit me in the nick, won't you?"

"You mean the psychiatric hospital. Actually it suits you. Looks great with the beard," said Cesare.

Beard? He was right. His beard felt as long as Cesare's. What on earth was going on?

Lucrezia went into the kitchen and came out with a pair of scissors. "Sit down, you two. Who's first for trimming? Look, you can't go down looking like Moses. Frighten the servants. What style would you like, Justin? Shakespeare? Van Dyke? Or just the latest modern?"

"A clean shave, please," said Justin. "I'd look idiotic with a beard."

"Look, I'm not a barber and there aren't any razors up here, so just keep still so I don't cut your ears off."

"I'll do my own. Speed things up. A Van Dyke, maybe," said Cesare. "Is there a mirror around?"

"Just the bathroom one."

"Oh, very dashing!" purred Lucrezia when he returned. "A real Don Juan. How's this for Justin?

"Yes, you look better with half your face covered up," grinned Cesare. "Now, how are we going to cover up the sexy bits so we don't cause a riot? Any ideas? Bit hot for travel rugs."

Lucrezia searched the cupboards and drawers.

"You can make yourself a Roman tunic out of these." She threw Cesare two small table cloths. He tied two pairs of corners together and slipped it over his head.

"Just needs a bit of rope or something." He went off searching.

"Sorry, no more tablecloths for you, Justin, but we can do something with these sheet sleeping bags. I can

46

make this into a frock, Greek style, if I can find any rope. We can make yours much shorter, man style, if I cut a hole half way along. Stick your head through here."

"We look complete idiots," grumbled Justin, when they had done all they could with ropes and dog leads.

"Tell that to the Greeks and Romans," said Lucrezia.

"And most people down the ages. It feels good to get back into comfortable togs like these. You must have worn stuff like this for thousands of years, Justin."

This was definitely another nightmare. Nothing else made sense. Why try to reason with it? He would wake up eventually.

The wolf pack was waiting for them outside the Refuge. Two animals were especially friendly. Cesare and Lucrezia sat down on the grass and gave them a good hug.

"Come meet Mom and Dad. This is my Sheba and that is Chez's Raoul. We've known them since they were tiny little mites. And you've got a fine family now, haven't you? Shake hands with Justin, Sheba."

Justin eyed the powerful creature dubiously. Under her bushy white eyebrows she had pie-bald eyes, one as blue as Cesare's and the other the turquoise of Lucrezia's. Her silver grey face was framed by a broad white ruff ending in a point, echoing her long pointed nose. She seemed to radiate goodwill.

"What a beautiful animal!" he breathed, fighting the urge to stroke her thick grey coat, speckled with cream. She was a wolf, after all. Who knows what she might do if she was hungry.

Raoul nudged Sheba aside and offered a paw as big and as long as a man's foot. He was a big, powerful animal with a tan-coloured coat, liberally sprinkled with black. The eyes under the black brows were as blue as Cesare's.

"Raoul, you are a very beautiful animal, but I'd hate to meet you in a dark, deserted alley."

"If you were alone in a dark alley you'd be very glad of his company," Cesare chuckled. "I certainly would."

Maybe this was Heaven after all, he thought, where even the wolves shook hands. Well, some of them did. He beckoned to the defeated wolf, lurking dejectedly behind the others. It half crawled towards him on its belly, giving its tail an apology for a wag. It cringed when he stroked it at first, then licked his hand. There was no doubt that Cesare could strike terror into anything daft enough to challenge him. Wise to keep that in mind.

"Well, Chez, the staff should be delighted with you. You fixed their wolf problem with your bare hands within minutes of arriving, while they can't cope even with dart guns. Are you sure you want those captured virgins? I'll make you a laurel wreath if you like."

"I'd rather you tended my itches as well as my scratches," he grinned.

He was certainly covered in scratches. Justin wondered if they would take as long to heal as his own wounds had done the week before. That should be interesting.

At first Justin was acutely aware of anything hard or prickly underfoot as they set off barefoot down the hill, but his squeamishness made him lag behind. The other two strode ahead confidently, so he took a deep breath and speeded up, determined to ignore the signals from below. It was only a nightmare, anyway.

"Should we tell them about the crash?" asked Lucrezia.

"I expect it will get into some newspaper or other. Everybody loves wallowing in plane crashes. And what do I tell Gino?"

"Well, since we are all three here it must have been a pretty nasty crash, nobody in hospital, all passengers thoroughly dead,."

"The helo was full of fuel, almost certainly exploded, went up in a fireball. There'd be nothing left of any passengers, even if they were human."

Justin's stomach turned over and the hair stood up on the back of his neck. Maybe this was not just another nightmare. He had been in a helicopter that seemed to be falling out of the sky, spinning him over and over like clothes in a washing machine. How had he got from that helicopter onto this hillside? Why did he feel uninjured, super-fit, in fact? That piece of metal machinery had hit him, gashed his shoulder; blood had soaked his shirt. He stopped to look at his shoulder. Perfect. Either this was the next world or this whole thing was a hallucination. Maybe he was in hospital, in a coma, hovering at death's door. His knees buckled and he slumped onto the grass, while the wolves danced around him, licking his face.

Lucrezia sat down beside him. "What's the matter? . Wolves bothering you? Shall I shoo them away?" She threw a stone and off they ran. "There, better now?"

"What's going on, Lucrezia? I can't make any sense of it. We were in a helicopter, but I can't remember getting out of it. Where are we now? It seemed like Paradise until the wolf attacked Cesare, and now a whole pack of them have turned into dogs. It's all crazy, like Alice in Wonderland."

"What's the matter with him?" Cesare sat down on the other side from Lucrezia.

"He's just a bit confused, I think," she said. "I don't think he does much helicopter flying, so maybe it was his first crash. This is Monrosso, Justin, our country estate. We made it. Thinking about it, shouting about it, worked. It was your idea, so we owe you. The house is down the hill. We'll get breakfast soon. That should cheer you up."

"Breakfast? We haven't had supper yet. Monrosso is up north near Lugano, isn't it? Last thing I remember we were over a hundred kilometres away from Lugano, and then we had another thirty kilometres to go by taxi, so you said. I can't remember any of that at all. Did I sleep through all of that? And what are we doing up on top of this hill? Did we land here instead of Lugano? "

"Justin, Sweetheart, we crashed that helicopter last night, just after we left Voghera. We all got burnt to a frazzle again. It's ridiculous, isn't it? That's two days in one week where we lasted only about twelve hours. It's getting as bad as some of the wars. Has the same thing been happening to you, or are you managing something more like a human lifetime normally?"

CHAPTER 5

PARADISE

Justin staggered down the hill, supported by the other two making sympathetic noises.

"I suppose, if you manage to lead a fairly quiet life, you don't get so much experience of dying. We've managed quite long lives sometimes," she said.

"Speak for yourself," said Cesare. "Try to turn myself in for a refit if I manage more than ten years at a stretch. Feel so much better afterwards. Feels great when you first wake up, don't you think?"

"Yes, it did," sighed Justin. Was this a migraine coming on? Stress always did that to him. Did you have to suffer migraines in Paradise? Was he coming out of a coma? Would it hurt like Hell when he was fully conscious?

"What do you think of the house?" asked Lucrezia.

There, ahead, stood a day-dream of a house. It was pale golden stone. Enormous sash windows overlooked a great expanse of open landscape, where two majestic cedars stretched out sombre branches. Water sparkled through the distant trees.

"Capability Brown did the landscape," said Lucrezia. "Nice man. We had awful trouble persuading him to find time to come over here. He said he hadn't quite finished landscaping England yet. He was worth it, wasn't he?"

The house was typical Palladian, an imposing central hub with Grand Order columns across the entrance steps, and a pair of matching wings.

"Palladio was such a copy-cat. He lifted lots of his designs from Vitruvius, but why not? Vitruvius has been dead since 15 AD and his work deserves reviving, doesn't it? Very harmonious, don't you think?"

"It's Paradise," said Justin. "How can you bear to tear yourselves away from it?"

"Well, don't you sometimes feel that fate has been too kind to you? That you ought to think of the poor humans for a change, see how they have to live. The poor things have only one life, and more often than not it's pretty horrible."

"Most of my life so far hasn't been up to much," muttered Justin. "Lots of work and worry. Everyone sneering, trying to keep me down. Scruffy homes."

"Your flat is gorgeous!" exclaimed Lucrezia.

"Only moved in a year ago. Rented a grotty flat in Rome till then. Owned an even grottier one in London. Done a lot of travelling, stayed in famous hotels, but had to work so hard there was precious little time to enjoy it."

Had it really been as bad as that, he wondered sheepishly. Or was it just the green-eyed monster talking? Everything he'd prided himself on achieving looked pathetic compared to this.

"You must have had lots of bad luck through the ages, then. I suppose you could lose everything each time you die if you don't have a partner to help sort things out. It's taken a lot of scheming to hang onto Monrosso but it's well worth the effort. But stop a minute. Here we are, blithely following our noses down to breakfast. How are we going to cover up our deaths this time? Any ideas, Justin?

It was all just some kind of nightmare, thought Justin, so what did it matter how ridiculous an idea sounded.

"I guess Cesare just jumped out of the plane at the last minute. There was no door on your side so you didn't have to force it open against the wind. You grabbed a prickly bush to drag yourself out of the river and got scratched to ribbons."

"Good man!" enthused Cesare. "Perfect! I'll go with that. But what about you two?"

"We met some friends at Voghera. The weather was terrible for flying, so they insisted we went home with them. Saw Chez crash and picked him up. Our clothes were all wet with the storm - and Chez, of course, got soaked in the river. Luckily they were having a Roman fancy dress party. Their staff hadn't finished washing our clothes when they gave us a lift home. Will that do?" asked Lucrezia.

"Well, we should try the usual first: 'Don't ask!' Usually goes down okay if we laugh when we say it," said Cesare. "Best say as little as you can - easier to change tack if it's not going down well. Don't you find that, Justin? We'll see better how things stand when I've talked to Gino."

"Better do that first. Put him out of his misery," said Lucrezia.

"Breakfast first. Let nothing come between a man and his breakfast," said Cesare. "Sneak in or brazen it out?"

"Not the front entrance anyway," said Lucrezia. "Don't want to face a reception committee dressed like this. Let's head for the back."

The wolves were bouncing around the courtyard behind the house, howling and barking like dogs with tonsillitis. A small, plump figure in white erupted from the building waving a frying pan. The wolves came galloping towards them. Cesare picked up a piece of fallen tree branch and flung it into the far distance.

Wonder if that would work for me, thought Justin, watching them tear off up the hill.

"Don't think Cook spotted us, but we obviously can't sneak in that way. Let's go to ground."

Cesare led them into a shrubbery and ducked into a grotto half hidden by creepers. There was daylight ahead but he swung away from the light and headed into a coal-black void. "We can risk lighting up now, Justin," he said quietly.

Gradually a dim golden light began to break up the darkness. Justin could make out the bulky shapes of rugged damp stone walls, and, as the light increased, festoons of spiders' webs. Tiny eyes shone briefly, then disappeared with a scuttling sound. Ugh!

Where were the lights, he wondered. Something was lighting up the tunnel, but what? The light seemed to be everywhere ahead, walls, roof, floor, all evenly lit.

"Clever lighting," he said. "What is it? Certainly not LEDs. Wrong colour, and no source, not hidden behind rocks or anything."

"Ouch!" said Lucrezia, bumping into him. "Light up, Justin."

"Sorry, seem to have lost my torch," he joked.

"Well, just glow; come on, do your bit."

He stopped and turned to peer at her in the darkness, but she was clearly visible, lit up like the walls. The hair on the back of his neck began to stir. It had become a familiar feeling of late.

"Glow!" she repeated. "Come on." The light around her began to brighten until he had to narrow his eyes.

"Steady on," he protested, "turn that down a bit."

"Well, then, do your bit. Light up so I can see you properly."

"Where's the switch?"

"Don't tell me you haven't discovered how to glow! Honestly, Justin! All you have to do is will yourself. Go on, make an effort."

Must be hallucinations. Must be sinking back into that coma again. Glow, you great wuss! You can do anything in a nightmare. Hey, look at that now!

"That's better!" exclaimed Lucrezia. "Don't want to stub my bare toes on you.   Now, let's catch up with Cesare."

He bumped into Cesare, too fascinated by the spectacle of his own glowing hands – and arms – and legs – to notice where he was heading. He felt a ridiculous urge to break into childish giggles. Look, I'm lit up like a Christmas tree!

There was a scraping noise as the wall ahead began to swing out towards them. "Steps now," warned Cesare.

He followed Cesare up to a landing at the top of the dusty stone steps. Light, a thin strip of white light this time, showed the faint outline of a door in the wall ahead.

"Now, listen carefully, Justin. This door leads into a staff staircase. Once out there we'll be on camera, so it's best to get a move on, straight to our suite; no stopping to admire the view. Okay?"

"Yes, but let's not be undignified about it. Surely the staff will want to have a good look at our fancy dress," said Lucrezia. "They'll rerun the tape. Better turn yourself off, now, Justin. Don't want to frighten them."

"How do I do that?" asked Justin.

"Oh, Justin, for Goodness sake! Just tell yourself to stop glowing."

It was easier said than done. Stop glowing, he willed himself silently. Stop glowing, you twat! Thank goodness the glow seemed to be fading. Pity, it was fun. Oh no! The glow was getting brighter by the second.

"Justin!" she wailed.

The door was wide open now. Ahead, was a spiral staircase, lit by a casement window. They let Lucrezia lead the way, leaving Cesare to close the door, which became completely invisible from that side.

She paused at the top of the stairs and put an eye to a lens in the door.

"I think the coast is clear. Stay close to me, Justin." she whispered.

She pushed open the door and led the way out onto a palatial landing that took his breath away. A pair of elegant, curved staircases swept down to a circular hall with a floor of inlaid marble, its swirling pattern radiating from the coat of arms on the central disc. Light streamed down from a huge glass dome. It may be an old house too, nearly three hundred years old, but what a difference from the B House! And it left his light, bright penthouse standing. Lucrezia tugged his arm and dragged him through the nearest door.

"Made it," said Cesare, closing the door. "Now, they'll be up here in minutes, so you two get dressed while I deal with the staff. Can you find him something to wear, Krish,?"

They heard a distant knocking. Cesare went through into the bathroom, dragged a dark red bathrobe off a hook, and strode off through another door, pulling it on as he went.

"Hello!" he bellowed, on route to open the door.

"My Lord, is that you?"

"Yes, hello, Fermi, great to see you again, looking very dignified, as always."

"It's a relief to see you, My Lord. All the staff were worried when you didn't arrive last night. And we heard rumours of a helicopter crash. They'll be thrilled to see you're safe. Our Lady Marchesa, is she in good health?"

"Yes, we're both as fit as fiddles. Sorry to hear we've given you a scare. Please convey our thanks to everyone for their kind concern. We'll do the rounds, say hello to everyone, once we get out of this ridiculous fancy dress."

"Was it a good party, My Lord?"

"A riot," laughed Cesare. "But not to be repeated, I think."

"What can we do for you, My Lord? Have you breakfasted this morning?"

"No, not yet. Breakfast would be much appreciated. We saw Cook on the warpath, fighting fit and furious, as ever. Is there a chance you might be able to sweet-talk her into putting that frying pan to better use? Our guest is English, Mr Justin Chase. You know, the computer genius. I expect he'd like the full English. Is that so, Justin?" he shouted. "You'd like the full English?"

"Why not?" said Lucrezia. "Yes?"

"Yes, sounds great, if it's not too much trouble."

"So, the full English for three, if you please."

"Shall we lay a buffet in the Morning Room, My Lord?"

"No, no, I don't think that will be necessary," Lucrezia called. "Just the standard English on a plate will be fine."

"Well, My Lord, shall I help you to dress or -?"

"Thank you, but I'm sure I can manage. Perhaps you would try your powers of persuasion on Cook. I think we're all ravenously hungry."

"Immediately, My Lord."

"Thank you, Fermi."

He strode back into the dressing room. "Good, you've found some pants. Yes, those chinos are a bit tight for me, so they may be fine for you. Here's the shirts, take your pick. There's loads of sandals and loafers there, look, so just try a few on."

"Look, Chez, I've found these dark red underpants," said Lucrezia, "and a long sleeved red shirt so if you go on bleeding it won't show. Trousers will have to be dark navy. Shouldn't show the blood. No, you can't wear shorts with all those scratches. Get in the shower now and clean the wolf off. I'll go find a frock and some

57

sandals, then I'll come and scrub your back. If you need the bathroom, Justin, there's mine on the other side of the suite. I'm going that way now, so follow me."

The Royal Suite in the Borgia House had been very grand indeed, but this was even grander. Like a golden wedding cake, everything was yellow and white and liberally enhanced with gleaming gold. The white and gold chairs and sofas in the sitting room sat on a thick and luscious circular pale yellow Savonnerie carpet with a soft red patterned border. Against the walls, delicately inlaid writing desks and console tables bordered with brass patterns stood on gleaming polished floorboards. Yellow walls and white ceilings were linked by ornate white cornices, touched with gold. The huge windows were dressed with curtains of white brocade, lavishly patterned with gold.

In the bedroom the huge and cosy-looking bed had a white and gold ornate head-board, heaped up with pillows matching the curtains and the fringed bed cover. This was surely Paradise. Nothing he'd ever seen on Earth was as bright and as beautiful as this.

Cesare seemed to have come a very long way from a war tent, from dossing down in a cave, on the run, tearing off flesh with his teeth to satisfy his hunger. The contrast was ludicrous. Did he really sleep in this fabulous bed, this 'thick-headed' man she had described as 'barely house-trained'? Well, she had told him lots of lies, hadn't she? Did she ever tell the truth?

Lucrezia's bathroom was both elegant and practical, with lots of neatly built-in drawers and mirrors, all white and timeless. Beyond the bathroom he could see her dressing room, apparently just the same as Cesare's, lined with pale wood fitted wardrobes with mirrored doors. No quarrelling in this house about encroaching on each other's wardrobe space, as he'd often heard his parents doing.

58

He still hadn't found any footwear. Better find Cesare's dressing room again. He crossed the suite again to Cesare's bathroom, a replica of Lucrezia's.

"There's no place like home - " Cesare was exercising that beautiful bass voice again.

"Need a back scrubber?" he shouted, over the noise of the shower.

"Yes, thanks, Pal." Cesare handed him the loofah.

"Last time I did this I'd just dug you out of the rubble. You were covered in cuts and scratches then as well. Doesn't it hurt when I scrub you?"

"Yes," said Cesare.

Justin stopped in consternation, "Sorry, didn't realise. You don't even flinch."

"What's the point of flinching? Does no good. If they're torturing you why give them the satisfaction of knowing they're hurting you? It's okay, carry on. These scratches need a good scrub. Make them bleed, if you can, wash the dirt out."

"Shouldn't you have a tetanus injection, or something? Great deep scratches like these must be dangerous."

"Well, over the last twelve thousand years or so we must have grown immune to nearly everything, but if something does lay me out you'll just have to do without me for an evening, won't you? I'm sure you'll be able to find some way of amusing yourselves."

"Can't get my head around what it must feel like to be immortal. You must have a completely different attitude to risk. And to screwing things up. There's always going to be another day when you can try again. You can even wait till everyone who saw you screw up's dead."

"Thanks, Justin. Saved me a job," said Lucrezia, breezing in. "I'm ready now, so what about you? Shoes? Why don't you raid Chez's wardrobes while I dry him."

The shoes were all lined up along the bottoms of the wardrobes. It was a good excuse to open all the doors. One housed only evening wear. There were two tail-coats, one red with brass buttons; three black dinner jackets, one of the finest silk, with different lapels; three white ones, one with three buttons, two with two; one in claret velvet, one dark green. Claret velvet with fancy frogging? A smoking jacket?

The wardrobe in the corner? Mid Highwayman Period! Oh my God! A fine knee-length cream jacket, patterned with gold and silver thread, with a matching long waistcoat. Pale cream doeskin breeches. A white shirt covered in frills, and shoes with high heels and big buckles. Crikey Moses! Imagine the tough warrior in fanciful gear like this! With rouged cheeks and a beauty spot! What a sight that would be! Was this fancy dress for parties? He felt the fabrics. No party costumiers used fabrics of such rich quality. These must be heirlooms, like Lucrezia's beautiful Renaissance dress he'd been crazy enough to try to destroy. He must have worn these once for real.

"I'm tempted to throw a fancy dress party just to see him wearing that again," said Lucrezia,

"Cuffs are a damned nuisance: trail into your food. You can eat seconds out of all those wrist frills," said Cesare, throwing his towel to Lucrezia and reaching for the blood red underpants. "No, don't like this red shirt. Horrible neckline. This navy one won't show the blood. Mmm, a bit dismal. Much better with cream chinos, but no, they'd show the blood. Now, where are we having breakfast?"

"I've told Maria we'd like it out on the porch. She says the table is still out there. It's not too far to carry it all,"

"You two set off while I finish dressing, speed things up a bit," he urged. "I'm ravenous."

"I think I need a course in envy management," said Justin, as they walked down the elegant staircase. "This is the nearest thing to Paradise I've ever seen. You don't know how lucky you are."

"Oh, yes we do. We love it to bits."

"But you've hardly been near the place for - how long? About twelve years?"

"We had quite long holidays from university, and we've tried to spend a few weeks here most years. We use Skype a lot to talk to everyone face to face. If we don't have to go back to the B House any more we can spend even more time here."

"When are you going back to the field hospital?"

"No idea. Can't go back to Fallujah. After all the fighting everyone's either dead or disappeared. The monsters tried to claim it was an American rocket that destroyed the hospital. I saw the fanatics with my own eyes, butchering the wounded, smashing up equipment. Of course, when I tried to remonstrate, they made short shrift of me. I'm not sure I can face any more. Not just yet. It's never-ending. Endless streams of people, all deliberately smashed up by other humans. It makes me want to run around screaming, to get off this hideous world somehow. Doubt if I can stay sane if I see any more of it. But it seems so greedy and heartless, swanning around here, in this wonderful place, while so many people are suffering."

"Well, any rational person's bound to agree you've done more than enough for a while. You surely deserve a good long rest at home."

"I'll have to see how Chez feels. I can't sit at home while he puts himself through hell. Is this a good place for breakfast, do you think?"

"Wow!" breathed Justin. Was there a good synonym for Paradise? That word must be getting worn with overuse.

The table sat within the colonnade, sheltered from the dazzling morning sun. There was no breeze yet to harry the white tablecloth, set with gleaming silver, and white and gold Meissen china. Lucrezia poured orange juice into three gold patterned Venetian glasses, then sat back to peruse the view, the elegant simplicity of the Capability Brown landscape. "Who said, 'Less is more'?"

"Mies van der Rohe, the Modernist architect. My kitchen designer's excuse for charging an obscene amount of money for planning rooms with almost nothing in them," Justin laughed.

"Well, Mr Brown had to put an immense amount of effort into creating this. He had the workmen strip out all the trees and bushes and move tons and tons of earth to make this great sweep of lawn. There were lots of little hills to level and hollows to fill in. Luckily he didn't have to dig a lake for us, because we thought Lake Como was quite enough to make a pretty picture."

"You were right there," breathed Justin, eyeing the water glittering through the trees. "Does your land stretch right down to the lake?"

"No, our boundary is the road around the lake, some way above the water. The Estate Manager's house is down near the road, guarding our entrance drive.

"Did Brown plant those two huge dark trees?"

"The cedars? Cedars of Lebanon. Yes, they're fabulous, aren't they? They're such a strange shape: the branches look as if they're floating in water. There are some Blue Cedars behind the house. Maybe you spotted them on the way down. Ah, here comes the Full English. Come on, Cesare! There's food here. What's keeping you?"

Cesare was right behind the two servant girls in their white frilly aprons, chivvying them along. "I don't know which is the nicer sight, you two or the breakfast," he grinned, as he sat down and picked up his napkin. "I remember you, Heidi. Lovely to see you again, looking so bright and perky, but I don't think I've had the pleasure of meeting you, my dear, have I?"

"I'm Judy, My Lord," she simpered, bobbing a curtsey. "I'm new. I've only been here three weeks."

"Well, I hope you're enjoying your new job. I expect Heidi's teaching you the ropes. If you listen to her you can't go wrong. She's first rate, aren't you, Heidi?"

Now the other girl was blushing too. "I do my best, My Lord. It's such a pleasure to work here."

"I shall tell Mrs Lepanto and Mr Fermi what you said," said Lucrezia. "They'll be delighted to hear you're enjoying your work. Thank you both."

Justin gave them a nod and a smile as they bobbed another curtsey and hurried away. Was that the right thing to do, he wondered. Should he have said something? If so, what? How was the guest of a marquis expected to behave? How did it feel to own all this?

"What strange sheep!" he said. "Great long necks. Is that really big one a llama?"

"Mmm.The small ones are alpacas.The wool they grow is more expensive even than cashmere. We sell it to the couturiers in Milan. Oh, those dratted wolves again!"

The wolves were playing sheep-dog, herding the alpacas into a huddle. The poor things squealed in protest. Suddenly there was a loud braying noise. A huge donkey charged out of the trees like a raging bull, scattering the yelping pack in every direction, gnashing its teeth at their rumps. The llama joined in, aiming kicks at the fleeing wolves.

"I don't believe it!" Justin protested. "A whole pack of wolves driven off by a donkey and a llama. This really is a nightmare."

"Well, it's fun to see it working, isn't it? We got the idea from the Internet. It's an American Mammoth donkey. Farmers in the States use them in wolf country, to protect their sheep and goats."

"How many donkeys does it take to do the job?"

"You mustn't have more than one, or they go off together and forget about the alpacas. She guards them because she's nothing else to call a family. Same with the llama."

"Poor lonely creatures," said Lucrezia.

"We'll let you take them a carrot," grinned Cesare. "What's the matter now?"

Shouts and yells were coming from somewhere in the near distance.

"Goats in the garden?" said Justin. What's that mean?"

"Red alert! All hands to the pumps," laughed Cesare. "Lucky we've finished breakfast. Come on, Justin. It's an emergency!"

CHAPTER 6

RUMPLESTILTSKIN

Cesare led them at a cracking pace, leaping down the grand entrance steps like a dancer, then running along the front of one wing and around the corner. They were faced by a stone wall at least four metres tall, broken by an open door, blocked by a young man with a flushed face.

"It's Rumpelstiltskin, My Lord. Cheeky little blighter! I've got to stay here to stop any more getting in." He backed out of the way to let them in. "He was in the asparagus bed when I last got a sight of him."

"He's got expensive tastes," laughed Cesare. "Right, Varmint, where you hiding?"

"Looks as if Cook's spotted him," said Lucrezia. "Over there."

Cook was tiptoeing along the path, frying pan aloft like a banner. Suddenly she lurched forward and took aim with the pan. There was a flurry of movement and the goat fled, dancing along the path, then stopped and turned to stare at them, a large pink flower in its teeth.

They were all flushed and panting by the time the goat finally grew bored with playing tig with a garden full of humans, all leaping about laughing and cursing. It danced a side-to-side jig with the gatekeeper, butted him in the knees and barged out. Then it turned to give the pursuers a last defiant look, the pink flower still between its teeth.

"That's goat power for you," laughed Cesare. "One small goat equals total mayhem. Imagine if a herd of them got in."

"There'd be nothing left for dinner, certainly," puffed

Cook, red face oozing perspiration. "You'd enjoy roast goat for dinner, wouldn't you, My Lord? That little demon's worse than all the rest of them put together. All that asparagus he's eaten should make him really tasty, don't you think?"

The goat was still staring at them quizzically, then suddenly appeared to get the message and scampered off to safety up the hill.

"Little devil!" shouted Cook, shaking her fist at it. "You're living on borrowed time."

"Cookie, my love, that breakfast was superb. Even my English friend was impressed, and he cooks up a first rate one himself," boomed Cesare.

"It was superb," Justin confirmed."Thank you so much."

"Well, we strive to please," she simpered. "Must get back to work."

"How do the goats get on with the wolves?" asked Justin.

"They'd fight a good fight if they had to, but they've got their heads screwed on properly. They rush to the cattle for protection. We've got Highland cattle, and they really hate the wolves, go for them like the donkey. And they've got big horns. Could gore a wolf pretty badly."

"So, the wolves have quite a hard time, then. Why is everyone so scared of them? I've seen lots of reports of farmers complaining about their protected status and trying to shoot them out of existence."

"Well, they are dangerous, of course. You've got to put the right kind of livestock together, as we've tried to do. And our wolves aren't hungry. Keep this to yourself — we'd get a hefty fine - but we eat the best parts of each carcase and feed them all the rest. They do kill the odd animal occasionally. You have to accept that. They almost always pick off a sickly one so it saves vet's bills, and we wouldn't be too keen to eat a sick one anyway."

"I think the staff are getting a bit rattled, now all the cubs are grown up. We surely can't cope with any more. Do you think we'd be allowed to stop them breeding, or maybe try to find two or three of them another home?"

"'Spose you're right. Better talk to the vet. Must have a word with the men, since we're here."

Justin took himself off on a tour of the huge walled garden while the Master and Mistress exchanged pleasantries with the four perspiring men who came towards them. That little goat had managed to keep eight humans on the hop. Full marks, Rumpelstiltskin!

The goat had done well to find the flower, for it was not a flower garden. Trees spread their branches laden with fruit along the walls; soft fruit weighed down rows of bushes, and a long raised bar was awash with strawberries. Courgettes, beans, carrots and a host of other plants he couldn't put names to seemed to be thriving, sheltered by the high stone walls. There were a few flowers too, big showy ones that might decorate large public rooms. This was not a pleasure garden, though it was a joy to look at: it was a kind of factory, a food factory. Maybe they had a flowery pleasure garden somewhere else. But wouldn't the goats soon chomp it up, and the cows, and the alpacas and the donkey? No. Pleasure gardens were probably off the menu, he decided, as he rejoined the others. It was a farm, after all.

"What?" Cesare exclaimed. "When was this?"

"Yesterday afternoon," said one of the men. "Must have got cracking as soon as he heard you were coming. I asked him what he was up to and he said he was just taking his sister a few things for her new house. The wife was going crazy inside, banging doors, flying out stuffing stuff into the van. Drove off like a lunatic and left the door wide open."

"Jeeez!" breathed Cesare. "Better go take a look."

"I've got the Jeep around the corner, My Lord. Shall I drive you down there now?" suggested a grim-faced retainer.

"What's the matter?" asked Lucrezia.

"Albertelli. Seems to have done a bunk. Yesterday afternoon, when he heard we were coming."

"I don't believe it. Why on earth would he do that?"

"Well, My Lady, maybe it had something to do with not paying our wages - " ventured one man cautiously.

"Not paying – how long since you were paid?"

"Six weeks."

"Oh, my - ! How on earth are you managing?" she gasped.

"With difficulty, My Lady, but what could we do? He came up with loads of lame excuses, then said if we didn't stop harassing him he'd tell the police we'd been up to something nasty, then he shut the door on us. We'd just decided to go have it out with him when Mr Romanov told us you were coming home. We knew you'd see us right."

"We certainly will," said Cesare. "Yes, we'd better get down there right away, see what's been going on. Justin, sorry about all this, but, well, maybe you'd like to go explore the estate for a bit - "

"I'd rather go with you, if it's not too inconvenient. Don't fancy being a wolf's breakfast," he added, shamefacedly.

"You're very wise," said Lucrezia. "Come on, let's get in the jeep."

A formal drive lined with chestnut trees led down towards the water, and as the impressive wrought iron gates came into sight, they reached a fine stone house with a good view of the lake. If this house was a perk for their estate manager, he was a lucky man.

Cesare jumped out of the jeep and strode to the front

door, rapping on it loudly. Nothing stirred. He twisted the knob and the door swung open. He was already leaping back down the stairs when the other three walked in.

"Gone, definitely gone!" he announced. "And in a tearing hurry. Stuff thrown about everywhere. Just left the heavy furniture."

"I doubt he'll be back for that," said the driver grimly "Looking back, its just what we should have expected. We should have insisted on getting to the bottom of it earlier. Idiots, we were."

Justin followed Lucrezia into the office. Tensely she pulled open desk drawers and stirred the contents, then began work on one of the filing cabinets.

"Can I help?" he asked. "What are you looking for?"

"The accounts, the bills. The staff haven't been paid. He's been fobbing them off. It's obvious he's got into difficulties of some sort. It must be very serious because he couldn't face us. He's just run away. There should be some big blue ledgers somewhere, one showing staffing costs, employment taxes - Another one should have all the farm costs, special feed, vet's bills, machinery, EU subsidies. Then another with the house costs, property taxes. I insist we keep the old-fashioned ledgers up to date so we can't be caught out by computer viruses."

"Here," said Justin, extracting five ledgers from the other filing cabinet.

She reached for them with relief, then let out a big sigh. They were empty, unused.

Justin looked searchingly around the room. Where would I put them if they told a sorry tale? In the bin? No, easily retrieved. In the fire? He looked at the fire grate. Something had been burned, definitely paper, but it all proved beyond retrieval. Just a minute - here, stuffed down beside a chair - ledgers, blue ledgers.

"Here, Lucrezia!" he called.

She reached for them eagerly and quickly opened them. Then she laid them open on the table and pointed with pursed lips. The pages had all been ripped right out. The evidence had gone.

"Oh Lord!" she moaned. "How can I reconstruct this?"

"Don't you have accountants? They'll have copies."

"Vetriano's only deal with our personal finances and things in Rome. Albertelli was our accountant here, on the estate. I check the figures on the computer every Friday. They always look perfectly healthy."

"When did you last come here to check the books in person?"

She pursed her lips. "It's been ages. I used to check everything really carefully at first, but he was such an expert I felt I was insulting him − and then we were in the Middle East for ages. I've asked for this, haven't I? The figures he posted must be bogus, or there'd be no need for him to run away. This is scary."

"What is scary?" asked Cesare from the doorway.

"He's destroyed all the accounts. I've no idea what state our finances are in. Must be bad, or he wouldn't have run away, would he?"

"The betting shop's been after him - "

They all groaned.

"Get Denrico down here, Iska." said Cesare tersely. "This phone's dead and I don't have a mobile."

Iska got out his mobile, then went outside seeking better reception.

"What you doing, Justin?" asked Cesare.

"Tax bills here, telephone bills - all red reminders - here. Ones I don't understand here − maybe animal feed, farm machinery, you tell me."

"Tax bills!" Cesare scooped them up and scanned them quickly. "Grief! We could be in real trouble. Look at these, Krish."

She seized them hurriedly. "Oh my God! We're being charged for late payment as well. They're getting bigger every day. How on earth could we possibly owe so much? It's unbelievable."

"Lots of these bills are reminders," Justin pointed out. "This one's been outstanding for over a year."

"Let's try to get the whole lot on the table," said Lucrezia. "Could you search the other rooms, Chez, see if you can find any more, while Justin and I empty these drawers."

Justin worked through all the drawers, making a hurried allocation of bills to piles, while Lucrezia checked them and reallocated some. Thank God this isn't my company, he thought, but maybe I should get the auditors in pretty soon. Bedding your finance director might not be the best way to keep your finances in order!

CHAPTER 7

LET BATTLE COMMENCE

Justin glanced towards the window as a motorbike slewed to a halt. The rider looked suspiciously like the butler, minus his tail coat and white tie. After a hurried conversation, the two men trudged inside and looked around warily.

"Denrico! Iskander! What the hell has been going on here?" Cesare's voice was calm and level, but his anger was palpable.

"You tell me," said Denrico. "I gather he left in a hurry."

"And left us more than half a million down in unpaid taxes. And that's just the start of it. Just look at this lot!" He waved at the table, now piled high with unpaid bills.

Denrico raised his eyebrows and exchanged glances with Iskander.

"Oh my God!" he breathed. "Are you sure?"

Cesare waved the tax bills in his face, then handed them to him. "Look! How do you explain all this?"

"How do *we* explain it?" blustered Denrico. "You surely don't think we had anything to do with it, do you? We'd no idea, had we, Iska?"

Iska shook his head and put on a pained expression.

"No, I don't imagine for one moment that you had anything to do with it, but that's the point, isn't it? Why the hell did you let it happen?"

"Well, it was Albertelli's job, wasn't it? It's him you should be blaming, isn't it?"

"No, it bloody well wasn't! It was your job. You took him on and dumped it onto him so you could cop out, swan around amusing yourselves, pigging out on the fat of our land while he bled us dry."

"Look, you should see all these new EC regulations coming out every other day. He got us subsidies, you know. Then there's all the new tax rules and employment laws. Makes your eyes water."

"Lekrishta managed it perfectly well, left everything in perfect order for you. You couldn't be bothered to engage brain, could you? Couldn't even be bothered to check what he was up to. Just took him on so you could dump the dull stuff onto him and go off hang-gliding whenever the fancy took you."

"Look who's talking. You're just as keen. And don't you know Den won the Chamonix Cup last week? Fantastic performance. Best height and distance in the whole two weeks. Landed spot on target."

"But there's always loads of no-fly days when the weather's duff. You just couldn't be arsed, could you?"

"Well, you've no room to talk," said Iskander. "How much interest have you taken in this place in the last twelve years? Prancing around playing God, playing the great hero, while we slaved away keeping the place going for you - "

"Going down the drain, you mean. Thank you very much! Well, our reward for giving you free reign for the past twelve years will be the bailiffs throwing us out into the street."

"Such gratitude for twelve years' work. I've put loads of work into the gardens and the livestock," blustered Iska.

"You did that for pleasure, didn't you? Always mad on farming - "

"And I've done loads of repairs to the house, cut the electric bills right down," said Denrico.

"We still have electric bills? I thought you promised - "

"You want jam on it, don't you? Blood out of stones.

73

I'm packing my bags. You've seen the last of me," said Denrico.

"Yes, you free-loaders, that's just like you, isn't it. We give you free board and lodgings for twelve years and what's our reward? You've destroyed us. We'll be homeless, out on the street. Clear off, then, quick, or you and your bags'll be down in the lake."

"It's you that'll be down in the lake, and we'll hold you down," roared Iskander, squaring up to Cesare.

"Stoppit!" said Lucrezia loudly. "All of you. ChesRa, SHUT UP! NOW! Quiet! Thank you. Now, sit down quietly,"

"Not enough chairs," muttered Iskander.

"Well, fetch one or sit on the floor. NOW!" said Lucrezia.

Iska sat on the floor. Justin, feeling he needed to make himself scarce, did likewise. Cesare and Denrico threw themselves sulkily into armchairs. Slowly and calmly Lucrezia turned the desk chair around and lowered herself onto it like a throne.

"Now," she said solemnly, "we'll have no more of this. Men! When will you learn some sense? Got a problem? Have a fight. Not fit to run a picnic are you? From now on you take orders from me. Gottit?"

The three men glowered at her sulkily. They were remarkably similar, Justin realised. Strip them and they'd look like triplets, the same black wavy hair, though Denrico's was short and neat and Iska's a tousled mop. They were obviously related, and were treating each other as equals now. No more 'My Lord' this and that.

"Let's examine the situation calmly and rationally," said Lucrezia. "About twelve years ago we had a very important discussion. Right? Den, you'd been telling us about all the work you'd done in Somalia, rigging up windmills and solar cells. Iska, you'd been teaching

74

villagers in India to improve crop yield. You inspired us, made Chez and me want to do something for the poor humans as well. You encouraged us. You offered to run this estate for us, improve it, even, while we were away. I'm sure that letting us go bankrupt was the last thing you intended, but do you think we deserve to be abandoned to suffer for your negligence?"

There was a long silence.

"Can you guarantee he won't throw us in the lake?" muttered Denrico.

"ChezRa, will you undertake not to throw them in the lake?"

Cesare gave them a filthy look. "I will endeavour to restrain myself from chucking them in this lake."

"Not good enough," said Lucrezia sternly. "Try again."

"I vow not to throw them in this lake," growled Cesare.

"Will that do?" she asked the other two.

"Hmm," they said grudgingly.

"So, now, do you intend to abandon us or stay and help us stop this ship from sinking?"

"Well, it did happen on our watch and we just slept right through it, didn't we?" muttered Denrico. "Ought at least to help to bail it out, put up a jury-rig."

"Yep," said Iska. "Should stay and fight the bailiffs off."

"Will you swear to that?" she held out her hand.

Iska put his hand on hers and Den followed suit.

"ChezRa?" Chez put his hand on top.

"What about him?" asked Den. Justin hurriedly put his hand on theirs, then Lucrezia placed her other hand on top.

"Together we will save Monrosso," she said solemnly.

"Together we will save Monrosso, we swear," the four men intoned.

"Thank you, gentlemen. Now, battle plans. I must try to reconstruct our financial situation, see how much in debt we are. Chez, you talk to the tax man and our major creditors, try to get an exact figure of what they believe we owe them and when last they think we paid them. Den and Iska, you could lend us your expertise, help us get this estate to make as much money as possible. Are we getting the best returns from the farm, Iska? Den, could we make the House pay? Could we do weddings?"

"Don't see why not," said Denrico. "Cook would love it, and so would the girls. They're a bit bored with so little to do. They were thrilled to hear you were coming back. Talking about you having lots of parties."

"We'd need a licence to open to the public. Food inspectors would want to come in. May take ages to get things off the ground," said Lucrezia."Could you look into that, Chez? Get the necessary forms."

"I can help fix that," said Den. "I've a friend runs a cafe in the village. Helped her get ready for the inspection, met the inspector. I could try to persuade him to come here quickly."

"The local council have a meeting of the licensing committee next Wednesday. My pal on the committee would sneak it onto the agenda, I guess," said Iska.

"So, you all think we could start advertising pretty soon? Who's good at adverts?"

"Me," said Justin. "Have to concoct them all the time to sell my stuff."

"Great!" said Lucrezia. "Could you talk to Den and Cook and Mrs Lepanto about what we could offer and when? Iska, could you talk to them about what food you could provide. Do you think people would be happy eating alpaca, or will we have to stick to beef and chicken? I suppose we may have to get some sheep, though how they'd go down with the wolves I've no idea."

"Might go straight down their throats," grinned Iskander. "Could we entice more people to eat goat?"

"I'll have a crack at that, but you'll have to promise not to put Rumpelstiltskin on the menu," said Justin.

"Now," said Lucrezia,"since the phone is cut off here it's presumably off now everywhere on the estate, and until we know where we stand I daren't pay any bills. Justin, you're the computer whizz. Is there any way we could get onto the Net without a Wi-Fi hub? And if anyone has any old spare phones or computers we need to borrow them. We lost everything in the fire in the B House."

"Might be able to help there, if everyone keeps mum," said Justin.

"We'll talk about that," said Denrico. "I've done a few pirate rigs in darkest Africa. We should get onto that right away. Top priority."

"Absolutely," said Lucrezia. "So, anything else while we're all together? No? Well, then, Operation Save Monrosso. Let battle commence!"

CHAPTER 8

REMEMBER THE SKUNDUN RAID?

"Hello, Monrosso House. Can I help you?"

"There you are," said Justin, "we've done it."

"Can I help you, Sir?"

"Sorry to disturb you, Mrs Lepanto. It's Justin Chase, here in the dairy. We're trying to rig up a private phone system around the estate and it seems to be working at last, thank goodness. You can ring any of us anywhere on the estate now, we hope."

"Can we ring outside numbers, Mr Chase? Our mobiles have stopped working up here at the House."

"Sorry, no. Not on these old radio phones. Completely different system, very short range. Your mobiles might work down close to the main road. We'll get onto that problem next. Goodbye."

"Where's Chez got to? He has some sort of military internet gadget, hasn't he?" said Den. "Works anywhere."

"He's gone down to the Lodge with my mobile." said Iska. "Said he had to ring his pal about the helicopter crash last night."

"What helicopter crash?" asked Den.

"Search me!" said Iska.

Justin changed the subject. "I expect his military gadget got burned in the B House fire, but maybe if he could explain how it worked - "

"We won't bother him then. I had some pretty clever stuff out in Africa. I'll go see if I can find any of it. Might work here, with a bit of twiddling. Best bet for computers now is the manager's lodge as it's near the road. Might be able to tap into some neighbour's Wifi from there.

78

Most people don't bother with a password. There should be a phone mast for mobiles not too far away as well."

Den and Iska brought their laptops down to the manager's house. They could no longer get online up at the Great House now its Wifi hub had been cut off. By lunch time their pirate networks were working, now and then, until their unwitting neighbours switched off their unprotected Wifi hubs, generating howls of dismay from the pirates.

"So, what's next?" asked Iska.

"How near are we to being able to drop off the electricity grid?" asked Cesare. "You used to say you could make us self-sufficient pretty quickly, Den."

"I've got all the stuff. Just needs installing. Bit hot for roof work, but we could have a siesta, then tackle it this evening if you like."

"Let's do that," said Cesare grimly. "I'm amazed they haven't cut the power off already. We haven't paid a bill in ages."

"The dairy's been off the grid for years, and the manager's lodge. And all the cottages and outbuildings. There's just the Great House left to finish. It's so big it's a bit daunting."

"So is having no power in the House," said Cesare. "Five o'clock start?"

"Slave driver! Okay, okay, keep your beard on."

Fortunately the roofs of the Great House were typical Palladian style: gently sloping, covered with light grey lead and hidden from below by a decorative stone fence around the edges - thank goodness, thought Justin. If he did lose his footing he couldn't slide off. This was fun, probably more to his taste than the Himalayas, he smiled to himself. Stately home roof-walking was something he'd never done before. It was the perfect place to practise, where it seemed he couldn't go wrong. He forgot about the solar panels as he dodged around the ornamental chimney stacks, exploring the novel space.

The view was mind-blowing, the stuff of postcards and paintings. If only he could bring up a chair and sit and stare at the boats zig-zaging across the lake, the elegant Art Nouveaux hotels hugging the shores, the mountains, the houses dotted way up among the trees. No wonder so many celebrities had bought houses nearby.

"Justin, come grab one end of this!" shouted Denrico.

Heyho, no peace for the wicked. Cesare and Iska were already positioning a solar panel under Den's direction.

After an hour or so Lucrezia came out of the roof door with cans of low alcohol beer and buns.

"What a lovely sight!" she said. "Nice to see men doing what men are made for. You'll deserve a good supper after this. It's goat casserole."

"Not Rumpelstiltskin!" wailed Justin.

"It's out of the freezer, don't fret," she laughed. She sat down on the warm leaded roof and handed out the beer.

Denrico had been frowning at Justin for quite some time. It was becoming disconcerting, so he put on a questioning look and stared back.

"It's no good, I can't place you," said Den. "What was your name originally?"

"Justin Chase. I dropped the 'H' to make it catchy, means something like 'per sicurezza, or 'nel caso in cui' in Italian."

"Yes, I get that, but what was your name in the beginning, way back?"

"Justin Chase."

"Nobody had two names in Mrusha, did they? What was your name in Mrusha?"

"Justin Chase is on my birth certificate. I don't know anything about this Mrusha."

"You've forgotten Mrusha! Well, it was a heck of a long time ago."

"You can't have forgotten Mrusha!" exclaimed Iska. "Were you a lazy mozzy meal or a rock ape?"

"He means did you live next to the lake among the skeeters or up on the outcrop," Lucrezia explained. "Did you carry your water all the way home, or just put up with the bites to save carrying the water?"

"You've lost me," said Justin. "Haven't a clue what you're talking about."

Cesare was sunbathing with his eyes shut so Iska gave him a kick, then whispered in his ear.

"No, I can't place him either," muttered Cesare, "but he's definitely immortal. We all three burned to a frazzle in that old helo last night. There's no way he could have survived. We were all tangled up together when we hit the ground. Then we all three woke up here together, by the Refuge, all picture perfect."

"You must remember the Skundun raid," exclaimed Iska. "Nobody could possibly forget the Skundun raid!"

"Sorry," said Justin, with a helpless gesture. "Remind me about it."

"Mmm, well, we thought they were great lumbering dopes, but they sussed us out, didn't they? Waited till we'd zonked ourselves out celebrating Rockon, then came creeping in about three in the morning; killed every man, woman and child. Then the crazy devils burned down all our houses and the food stores and even the corn. What on earth for?"

"You must remember the shock we got when we woke up next morning: everybody else dead, hacked and burned, and us so perfect, not a scratch," said Lucrezia. "And we could all remember being killed, hacked and bashed and burned. Why had we come back to life? Why were we so perfect? Not a scratch. Surely nobody could forget all that. We thought at first we'd gone to Heaven – till we looked at Mrusha."

"What was left of it, you mean. Nothing but ashes and mutilated bodies," said Den.

What could he say? That apologetic shrug would have to suffice.

The other four continued to frown and look pensive, so it was a relief when they got back to work. At last, when it was too dark to see, they went down to the kitchen in their sweaty work clothes, speculating about Cook's goat casserole.

It seemed that Cesare was as economical with the truth as Lucrezia. So much for not barging into the kitchens to avoid frightening the servants! They all barged in, washed their dirty hands in the wolf sink, and flung themselves onto the chairs surrounding the big wooden kitchen table.

"Here you are," enthused Cook. "Don't you dare tell me this is not the best goat casserole you ever tasted."

"Cookie, my love," grinned Cesare, "we wouldn't dare. Come on, dole it out. I claim the biggest plateful as it's my house. Now, sit down, Cookie, and show us how to eat it. Then we'll know you haven't poisoned it," he added sotto voce.

Everybody laughed as Cook took a threatening step towards her favourite frying pan, then, smiling broadly, plonked herself at the table.

Lucrezia, you fraud! thought Justin, looking around the kitchen as he ate the tasty stew. It had exactly the same cupboards and drawers as she'd chosen for their house in Rome. No wonder she'd been good at choosing something suitable for the servants! And that marvellous steam oven he thought he'd talked her into. There it was, large as life. He cringed at the memory of how he'd underestimated her.

Listening to the conversation, he concluded that, although their relationship with the cook was very relaxed and friendly, she was probably not a fellow immortal. They were editing their conversation to avoid anything that could be construed as unnatural.

"Cook, darling, that was perfect, as always," said Lucrezia, giving her a little hug. "Now, why don't we take a drink outside so Cook can clear up and go to bed."

"Goodnight, Cookie, my love, Great supper. Perfect homecoming." Cesare picked up his glass and a bottle of Monrosso Rosso and led the way through into the grand hallway and out onto the colonnade.

They spread out the chairs from the breakfast table so they could all admire the view. The moon had not yet risen, so the vast lawn was a sea of blackness, topped by the lake, glittering darkly in the lights around its shores. A view to die for, thought Justin. Imagine losing all this beauty. Imagine the bailiffs driving you out of this Garden of Eden. They had to save Monrosso. Yes, he had pledged himself to do that, but only under duress. Now he needed no forcing. It's not mine, he reminded himself, but his chance of ever owning anything like this himself seemed vanishingly slim. This was the nearest he was ever likely to get, so banish the green-eyed monster. Think yourself lucky to be involved. Try to find some way to earn your place in these people's lives.

"So, Justin, you insist you didn't grow up in Mrusha," said Den, "so where, pray, did you originate? So far you are the first non-Mrushan Immortal we've ever met. Thought we thirty were the only ones in existence."

"Thirty!" Justin exclaimed. "Where are the other twenty six?"

"Goodness knows! Haven't set eyes on most of them for thousands of years. Some lucky sods might have found a way to dispose of themselves: who knows?" They all gabbled at once.

"So, you four may be the only ones still in existence, as far as you know?" said Justin.

"And now we are five," said Lucrezia. "So, tell us about yourself. Where were you born and how long have you existed?"

Justin sighed. "Well, I first saw the light of day in Cleckhuddersfax twenty-nine years ago."

"Yes?" said Lucrezia, "and before that?"

Justin had the Italian helpless gesture off pat now. It would have to do.

"Are you trying to tell us that yesterday evening was the first time you ever died and this morning was the first time you ever resurrected?" asked Cesare forcefully.

"'Fraid so," said Justin, feeling somewhat inadequate.

"Jeez!" breathed Cesare. "Somebody's given you immortality treatment? Who? Where? How did it work?"

"No!" exclaimed Justin, "certainly not. I told you I'm dead against creating any more Immortals. Imagine people like Hitler and Gaddafi living forever!

"So how do you explain yourself?" demanded Cesare.

"How the hell can I explain myself? If this isn't just a hallucination I haven't a clue how I've got myself into this mess. Must have caught it from you two, like Ebola,"

"Maybe he's not immortal at all," said Iska. "We should check him out. Kill him now and see if he wakes up tomorrow; then we'll be sure, one way or the other."

"Hey, that's not cricket!" squawked Justin. "You say you've existed for maybe twelve thousand years. I'm only

84

twenty nine. Give me a break. I'd like to live a little. I'm your guest. What about the ancient rules of hospitality?"

"Yes, leave him alone," said Lucrezia. "He's been very kind to me. Took me in when I needed help, so I owe him at least a holiday, whether he's immortal or not. He's my guest, so hands off him, yes?"

"So, welcome to our humble abode, wandering stranger," grinned Cesare. "Happy holiday!" He raised his glass, and the other three Immortals followed suit.

Work on the roof resumed on Sunday morning and again on Sunday evening. By the time the sun went down the job was done, tested and working.

"They can cut us off now: we're afloat at last. How about that?" Den gloated. "No more electric bills for you. Now, before we relax and assume we've done, is there anything we've missed?"

"We could do the Refuge. A supply up there might come in useful, power something. Could we electrify a wolf fence, do you think?"

"Good idea. We'll have to corral them if we're going to have weddings. I've got a couple of panels left. We'll do that first thing tomorrow before we run out of steam."

"It's Monday tomorrow so you can contact the tax man," said Lucrezia.

"Nobody's offered me a spare mobile yet," said Cesare. "Surely somebody must have an old one. I had several. Don't know if I dare use these wonky systems in case somebody might be able to eavesdrop. If people realise we're in trouble they won't want to deal with us. Maybe I'd better write, or try to find a proper phone down in the town. Have you got a fix yet on how much we're down?"

"Nearly a million and a half."

"Well, that shouldn't be much of a problem. You've

been selling big ticket armaments for months. Must be plenty in the bank," said Iska.

"That was all for funding field hospitals and refugee camps. We cleared that account last Tuesday morning, sent the money off before the place caught fire."

"What, cleared out everything? Crazy! Why ever?"

"Well, we raised the money for them, didn't we? What else could we do?" asked Cesare.

"Was it donations - did you rattle the collecting tins?"

"No, Chez raised it all himself, selling arms, and doing jobs for the Americans. So many casualties, so much suffering. We feel so guilty just abandoning them. The least we could do was give them the money. The figures Albertelli posted every week looked very healthy, so I thought Monrosso was doing really well. I was making a few thousand a week on futures and things so I guess we could scrape together about a quarter of a million from our personal accounts, but I suspect there may be a few more nasties here I haven't found yet. I can't see any way we could pay the tax bill, but maybe we can agree some kind of deal, show them how we're making Monrosso pay - "

"Then they'll just demand even more taxes, won't they?" said Justin.

"We'll have to sell something. Just let me think about it. We could let the Lodge out. We're not going to replace Albertelli, are we? That should bring in quite a bit. How can we make the farm pay more?"

"Couldn't we spin our own alpaca? The firm we sell the fleeces to sells the yarn for a price that makes your eyes water," suggested Iska

"We'd need to get a sample of their yarn to make sure ours was alright," said Lucrezia. "Maybe we could even try weaving it as well. We'd certainly need to find out what the fabric is meant to look like first, wouldn't we? Or try fancy weaves. See what impressed the designers."

"We could graze far more cattle, couldn't we? Our Highlands' meat is supposed to be very lean and healthy. We could probably sell lots of it in Milan. There's plenty of money sloshing around there," said Iska.

"We could certainly squeeze more work out of the dairy, more goat cheese and butter and preserves," said Den.

"And weren't we going to get some bees? You were making hives, and we planted loads of manuka bushes before we left," said Cesare.

"You could make a mint out of Manuka honey," enthused Justin. "There's a world shortage because it's so popular. I saw loads of manuka bushes flowering by the Refuge. If anybody near here keeps bees they're eating all your nectar. We ought to put a stop to that."

"The market stall, do we still do that on Saturdays?" asked Lucrezia.

"Oh, yes, the women enjoy that. Gets them into Como to do some shopping."

"I've not found the accounts for that. What kind of profit do we make?"

"Have to ask the women. Not much, after they've done the shopping. They split what's left among themselves."

"What!" exclaimed Lucrezia.

"Well, you surely don't expect them to do it for nothing, do you?"

"So," said Cesare, "they sell all our produce and turn it into shoes and handbags, do they?"

"No, course not. Well, they do sometimes do a bit of that, but, well, they're women, aren't they? But there's not much money left when they've been to the supermarket."

"So, we pay for their groceries as well?" asked Lucrezia.

"Well, fair does if they're not getting their wages," sighed Cesare.

"I don't know about that, but I do know that Maria Lepanto gives them a big long shopping list for the House, for the stuff we can't produce ourselves, like salt and coffee and detergents and loo rolls, you know. She's been hinting that the list seems to get longer and longer every week, but only the usual amount ends up in her cupboards. Maybe she's trying to tell us something," said Den.

"All our lovely staff are turning into thieves," wailed Lucrezia.

"Can you blame them? If we don't pay them they have to pay themselves in kind. We've driven them to it."

"I'm not sure if Monrosso's a communist commune or a feudal relic," said Justin. "Never realised before how much they might have in common."

"Jeez," groaned Cesare, "what have we done to this place?"

"Our stuff sells very well. They're all packed up by lunch time, apparently."

"Why don't they take lots more?"

"Well, they don't want left-overs cluttering up the van, and they've got to get to the supermarket. It closes quite early on Saturdays."

"Next Saturday we'll go along and help them. We can let them go to the supermarket and pick us up on the way back. That way we shouldn't spoil their fun," said Lucrezia.

"Have we tried selling to the shops?" asked Cesare.

"Oh, yes, we supply the delis in Bellaggio and Menaggio. They say it sells pretty well."

"Well, then, we need to step up production and find a lot more outlets," said Cesare.

"We could put up adverts for the weddings in the delis," suggested Denrico.

"If we're going into catering we'll need tables. There must be somewhere we can hire them till we're sure it's going to take off. Can't afford to waste money," said Cesare. "If only we knew how long we've got before the bailiffs turn up."

"Well, somebody's got to take us to court first and get a judgement, haven't they? Then, I suppose, we could lock the gates and fasten the wolves' food to the backs of them. How long before they bring the police with them, do you suppose?" asked Iska. "And then the army." He flexed his muscles and grinned.

"How are the wedding guests going to get in?" asked Den. "Fight their way through the army and the bailiffs?"

"And then get eaten by the wolves," spluttered Justin.

They all dissolved into hysterical laughter. It had a therapeutic effect, wiping out all the simmering resentments. When they finally calmed down they grinned at each other like the best of friends.

There was something to be said for a crisis like this, thought Justin. It certainly added a great deal of spice – of entertainment even - to life. Would "Save Monrosso from the Bailiffs" make a good computer game, he wondered, for players of a less bloodthirsty disposition?

For reasons he couldn't quite fathom, Justin found it hard to take this tragic situation seriously. Yes, his new friends would be broken-hearted to lose their beloved Monrosso, but he had never in his life met people who smelt less like losers. Everything about them, from their fine physique to their air of superiority and confidence, suggested winners, likely to come out on top in any situation, however dire.

Anyway, it was all just a nightmare, wasn't it?

CHAPTER 9

FOOD FOR THOUGHT

"First stop the bank," said Lucrezia, as they finished breakfast on Monday morning.

Here we go again, thought Justin, wryly. And this time he was not just a sympathetic onlooker. What can you do without a sou and not even a bank card to your name?

Cesare dropped them in the centre of Como near their respective banks and drove off to park the Jeep.

The bank proved every bit as frustrating as he expected. The manager's default setting was deep distrust. Well, yes, it was as well, really. It was proof that crooks would have trouble making a daylight raid on his accounts. At last, after endless questions, signatures and long spells of waiting, the manager came back all smiles, promised a new debit card in a couple of days, and authorised a two thousand Euros cash withdrawal on the spot. Justin walked out into the sunshine with a bulging pocket and a big grin.

Next stop, that other essential of Twenty-First Century life, a smart- phone.

He walked on past the information technology emporium and eventually found a quirky little shop in a side street. The old man behind the counter gave him a surly look.

"I don't suppose you have a Darkphone."

"Oh, you want a Darkphone, do you? Got some 'unusual' hobbies, have you?"

Justin put on a smirk. "Well, do you have a Darkphone or am I wasting our time?"

"Might have."

"So?"

The old man reached into a drawer and took out a smart black box.

Justin opened it, took out the phone and lit it up.

"How much?"

"Five hundred. It's a Darkphone Two, latest thing."

"How much for two?"

"Two! You starting a vicious circle?"

"Wouldn't you like to know?" smirked Justin. "Do you have another?"

"A thousand," said the old man, taking out another box.

"Nine hundred."

The old man pulled a face. "You'll need the right apps as well."

"You're right about that," said Justin. "Nine hundred for the pair." He offered the cash, and the old man reached out greedily.

Cesare and Lucrezia were already drinking cold lager when he arrived at their lunch rendezvous.

"Look what I've got," said Lucrezia, dragging a huge laptop out of a bag. "Amazingly cheap. I suppose big laptops are right out of fashion these days, but it should be perfect for the accounts."

"This phone was a bargain too," said Cesare. "Last year's model. Should do everything I want."

"Ah, but will it encrypt all your communications so no one can spy into your affairs?" asked Justin.

"Well, anything can be hacked these days, can't it?" said Cesare. "You just have to accept that, don't you?"

"Oh, no you don't," said Justin smugly. "Prezzy." He put one of the boxes on the table between them. "I gather house guests are expected to bring something, so better late than never."

"Oh, there was no need, but thank you all the same,"

said Lucrezia. She opened the box and took out the phone. "Very smart. What kind is it?"

"Darkphone," said Justin. "Something new, created for encryption. They've built their own network of Dark Net servers, and there's one near here in Switzerland. They're all guaranteed unhackable. You can phone your family without having to be careful what you say. You can send texts, send files and pictures, even have live video conferences, all absolutely secure."

"Well, that could be useful sometimes," said Cesare.

"It works normally on the Android system as well," said Justin. "It will do everything that one will and more."

"Ah, well then, it sounds like a winner." Cesare reached out for it, interested at last.

Thank goodness, thought Justin. It seemed he'd got it right. He needed them on his side to stop Iska killing him – just to see if he really was immortal.

"I'd better warn you that you might get some funny looks if you mention the Dark Net," said Justin. "Are you familiar with it?"

Lucrezia shook her head but Cesare gave him a guarded look.

"You are, Cesare, I expect."

"Mm, pretty nasty. Paedophiles, crooks and terrorists."

"And all the security services, CIA, MI5, your DIS, etcetera, - and most heads of state as well. The President of the USA doesn't want eavesdroppers when he's talking to the German Chancellor. The Dark Net's been around for nearly as long as the Internet. It's a huge matrix of secret servers most decent people have never heard of. There's been so much phone hacking recently that various firms are trying to provide Dark Net encryption for decent people who just don't want their affairs in all the gossip columns."

"Well, then, good luck to them," said Cesare.

The car park was relatively quiet, so Cesare began his phone battle with the tax officers while they sheltered from the heat under a tree. It was soon obvious that things were not going well. Keeping calm and polite was clearly demanding a very great deal of effort. Finally he ended the calls with a big sigh.

"Three weeks," he said. "That's when our case comes to court. They say they sent us several threatening letters. Did you find any of them?"

Lucrezia shook her head. "He must have destroyed them. Well, at least we know where we stand. We really are in dead schtuck."

As they walked over to the Jeep Cesare continued to check his emails. Suddenly he ground to a halt. Lucrezia leaned over to look at the screen and gasped.

"No!" she exclaimed loudly. "No, don't even think about it. Here, give it to me."

"Calm down, stop the fuss," he said, holding the phone where she couldn't reach it and closing it down. "Come on, let's go home."

"You promised to tell them you'd given up."

"You can drive," he said, opening the door for her. "Anyway, when have I had the chance? I've only had a phone since lunchtime. They're not exactly easy to contact."

"So, you can tell them now. You can use my new secret Darkphone."

"It needs setting up with the right apps," said Justin. "And your new laptop needs setting up as well. Can you do that yourself?"

"I'm not brilliant at things like that. I expect you could do it in your sleep, couldn't you?"

"I'll gladly do it for you, and the Darkphone as well."

"Oh, yes, please." She gave him one of her knee-weakening smiles.

Justin's innards began a wrestling match. Stop it: I love it! If only – but it's hopeless. What chance had he against a hunk like Cesare? And now there were three of him!

With Lucrezia fully occupied in the driving seat, the Hunk could safely continue dealing with his accumulation of emails in the back seat as they wound around the narrow twisting road above the lake shore.

At last she parked the jeep beside the manager's lodge and they took their purchases inside. Justin set to work immediately on the new laptop, introducing it to its printer and all the other useful devices that the manager had left behind. He set up the desktop with all the websites and apps she felt that she might need most often.

"Any tea around?" he asked "This is thirsty work."

"Da der!" crowed Lucrezia. "Look what I brought back." She produced a box of Yorkshire Tea bags and a tin of powdered milk. "I asked them if this tea came from Cleckhuddersfax and they were completely mystified."

"Yorkshire Tea, here in Como? Amazing! Still, I did read that it's selling like crazy in China, so why not in Como as well? Well, this machine is ready to go now, so why don't you Google 'Cleckhuddersfax' and see what you get. Probably tell you about the Treacle Mines too."

By the time he brought in the mugs of tea she was laughing and trying to sing a Treacle Mine song.

"Have you read about the Great Treacle Mine Disaster? Everybody got stuck to the walls."

"So, where were you born, Cleckheaton, Huddersfield or Halifax?" she laughed.

"A little village up in the Pennines that seemed a heck of a long way from each of them. Rained almost non-stop. A great place to grow up. Anywhere else you might end up is certain to be an improvement."

"Well," she sighed, "maybe we'll have to go live there if we're thrown out of here in three weeks' time. Property there must be cheap. Thanks very much for setting this up. I'd better put everything I can on here as quick as poss. Never say die. It might not be as bad as we think. I might be able to do something to fix it."

"I'll leave you in peace, then," he said. He picked up the third mug of tea and set off up the stairs to find Cesare, who had slipped out of the room with Den's laptop as soon as her back was turned.

Cesare was staring so intently at the screen he hardly noticed Justin. Google Earth. A gaggle of buildings.

"Waziristan?" he asked, plonking down the mug by Cesare's right hand.

"Syria. Raqqa."

"Oh dear!" said Justin.

"Alas, poor Raqqa. I knew it once - actually, three times, once in bad times, twice in good. Brought up a family there – twice. Great kids, two boys and one girl both times. Seems to be my standard issue."

"With Lucrezia?"

"No, no. Lost track of each other for centuries sometimes. When we did meet we were usually already married to other people."

"So she called you her brother?"

"Mmm. That way she wouldn't be stoned as an adultress. In Islamic circles a woman must always be accompanied by a male relative in public, so that gives us a perfect excuse for being together."

"I expect you've had lots children together."

"No, none. Not for want of trying. Just doesn't happen."

"So she can't have children, then?"

95

"She's had hundreds; we both have. But none together, sadly."

"So, the Americans want you to go to Raqqa?"

"Mmm."

"Why?"

"The usual. Somebody they can't reach with their Seals or even a drone."

"Does she usually kick up a fuss?"

"She doesn't usually know – not till I've already left home. Then she can't reach me till it's over."

"She doesn't approve of what you're doing, then?"

"It's not the killing she doesn't like. She's killed some monsters herself. It's the reprisals. You can't blame the bastards for wanting revenge, can you? It's usually hard to get away afterwards, and, in any case, it's better to stay and take the rap on the spot. If I did manage to get away they'd almost certainly torture loads of innocent people to death as a reprisal, wouldn't they? I keep telling her I'm used to it. I've got my own way of coping. I try to wind them up, get them so mad they bash me extra hard, finish me off quickly. I've been a warrior for thousands of years, been killed thousands of times. I can shut myself down, just think about being at home in bed – with Krish if I'm really lucky."

"Soldiering sounds a pretty uncomfortable life. Why do you do it? There must be easier ways to make a living."

"When you're built like me you don't get much choice. First sign of trouble and the shouting starts: 'ChesRa! ChesRa! Save us! Lead us to victory! What can you do? Run away and hide, and watch some idiot make a hash of it? Let the enemy slaughter your people, burn your home, drag your wife and kids away to slavery?"

"So, there's some advantage in being born a bit of a wimp," said Justin. "No one expects you to be a hero.

These assignments - do the Americans pay you? Will it be enough for the tax bills?"

Cesare took a sharp breath, shut his eyes and propped up his chin with his hands. "Not right, is it? Not right at all," he sighed.

"Your tea's going cold," said Justin. "And it's certainly the right stuff: Yorkshire Tea, grown on the moors above the Treacle Mines. You know, where the poor miners all got stuck to the walls."

"You daft 'aypeth!" grinned Cesare. "Aha, you thought I didn't know God's own county, didn't you. There's a great fish and chip shop in Slawit."

"Do you mean Slaithwaite?" asked Justin. "You're right: there is."

Cesare pushed back his chair. "We invited you to do a bit of mountain bashing, didn't we? And we've been working you to death instead. What about a hike before dinner?"

"I thought you'd never ask," said Justin.

Six in the evening was the perfect time for a stiff walk. They soon settled on a speed that suited them both – pretty fast. It was lucky Lucrezia had declined: she would have had to run to keep up.

What a place! Justin exulted. What a lake! So big, so full of life. So many little towns, all in colours as warm, as natural, as harmonious as a basket of home-made bread. Boats cutting furrows across the gleaming grey water strung them all together. So many elegant hotels, so many gracious houses, nestling among the trees, way up the lower slopes, and up on the skyline, the glorious mountains, in endless succession, green nearby, then tinged with blue in the distance. Imagine owning the right to walk here every day, drinking in this beauty. Imagine the bailiffs coming to take it away.

He began trying to calculate how much liquidity he had. Cash-flow was great, but so were his outgoings. He'd followed the experts' advice to put each windfall into property and residency rights in Rio and Singapore and Rome. Expensive homes demanded high outgoings. If he'd rented them out he couldn't live in them, could he? That little basement flat in Earl's Court wouldn't sell for much. As he'd lost interest in it he'd never even bothered to pay off the mortgage. Pity he'd spent so much on that ridiculous kitchen.

If he sold the business he'd be drowning in money. It was doing so well it might be time to do that, either sell to one of the giants or float on the stock market: either way he'd never need to work again. How long would that take to organise? Surely far longer than three weeks. Right now he could probably pull out little more than half a million Euros. And he had a feeling they would refuse to accept his help in any case.

Would they sell him the manager's lodge? Judging by the prices he'd seen in the house agents' windows in Como, it was probably worth up to a couple of million – chicken feed if he'd sold his firm, but as things stood he'd need another huge mortgage. Then, if he hit a down-turn, he could sink without trace, all his work for nothing.

But the Americans had paid Cesare millions before. He'd seen their accounts, nice round figures, a few million a time.

"You're not happy with the fee the Americans are offering this time?"

Cesare sighed. "Nothing wrong with the bounty. Could be nearly twice the normal. There are three they want to rub out, two killers and a talking monster. He's a preacher, very charismatic, preaches terror, torture, humiliating women. Wants to turn the world into the kind of hell you see on medieval paintings, everyone having limbs or heads chopped off, stabbed with pitchforks, boiled in oil, people running, screaming, devils chasing them. He's a murderous psychopath, just feeds on suffering and terror. He never risks his own skin - probably a snivelling coward. And he brainwashes families into turning their little kids into walking bombs, promises they'll go straight to Heaven – and their parents too."

Cesare speeded up, striding ahead as if trying to outrun the monster.

"Hey, steady on, slow down a bit," Justin protested. "You think this job might be too hard to crack?"

"No, could be a walk in the park. Know the city – it used to be home; speak the language. They've just announced they're opening a medical school. That's a perfect reason to go, isn't it? I'm coming back to my own country, disgusted with life in the West, among the Kuffars, the Rafidi, the enemies of Islam. No need to pose as an arms dealer, just offer to work at the medical school."

"But you're not happy about it, are you?   Why?"

"Killing to line my own pockets, feather my own nest? You think that's okay, do you?"

"Mmm. You told your brothers you'd raised all the money for the field hospitals, didn't you?"

"And I thought if I could get rid of the monsters I could stop their victims getting hurt, couldn't I? And I could use the money to treat the injured. Everybody gained."

"Well, obviously. Most rational people would agree you were doing a very good thing."

"Only it didn't work that way, did it? Kill one monster - "

"And another two pop up in his place," said Justin.

"Exactly."

They strode on to the top of the hill. There was a fabulous view of the whole estate, the beautiful house, the dairies, the refuge, the manager's lodge, the cedar trees. Monrosso had to be saved somehow.

"I can raise half a million for the Save Monrosso fund," said Justin. "Surely your brothers can chip in as well."

"My brothers wiped themselves out in the market crash in two thousand and eight. Finance has never been our strong point. I'm lost without Krish. She's the brains of this family. And thanks for the very generous offer. I really appreciate that, but I can't let you risk your business."

"Is there a risk you might let yourself in for some serious harm if you took on this job?" asked Justin.

"Apart from getting myself killed?" he grinned.

"Sounds ridiculous, doesn't it? You're sure you'd wake up next day?"

"As long as I was completely dead the day before. You've seen how it works. The worst thing would be ending up paralysed in hospital, with them doing their best to keep me alive. That's my nightmare. I'd just have to hope you lot would come and finish me off."

"Well, we'd certainly come looking if you didn't turn up. You could be sure of that."

"That's not the point though, is it? This is murder for personal gain. I have to live with myself. Maybe you can't understand that."

"Well, I suppose I can see what you're getting at."

"Better turn back now. Won't be very popular if we keep everyone waiting for dinner."

"So," said Justin, as they set off down the hill, "that monster will just carry on sending little kiddies off in bomb vests to blow up cafés and police stations."

"Two would pop up in his place. You said that yourself."

"Maybe I wasn't thinking straight. What if somebody had stopped Alexander the Great before he left

Macedonia, or Napoleon before he began taking over Europe? Two Alexanders? Two Napoleons? Two Hitlers? two Genghis Khans? Surely those people were unique, changed the course of history. Would events really have gone just the same way if they had been stopped in their tracks?" Think Savonarola, Oliver Cromwell. They changed the course of history, and when they died history changed right back to what it had been before, didn't it? All that killing and suffering for nothing. Surely some day the terrorists will run out of homicidal maniacs and calm down," said Justin.

CHAPTER 10

DECISIONS, DECISIONS

"Could we do a golden wedding lunch on Saturday?"

Everyone stopped talking and stared at Lucrezia.

"Which Saturday?" asked Den.

Lucrezia turned to look at the housekeeper who had just brought in a mouthwatering Raspberry Pavlova.

"This Saturday, Mr Fermi," said Mrs Lepanto.

Everybody's eyes widened.

"Are you serious?" asked Lucrezia.

"Well, My Lady, that's for you to decide, I think."

"Gracious!" said Lucrezia, "but I think the decision should rest with you and Mr Fermi. After all, it's you who will have to make it work. Of course I'll help as much as I can, but I've no experience of behind the scenes at parties. I'll need a lot of guidance from you both."

"We'll appoint you cashier and receptionist. They won't appreciate roast hedgehogs and pigeons," said Cesare.

Justin laughed wryly. So, she had used that same ploy on her husband as well as on himself – anything to wriggle out of doing the cooking. Naughty creature!

"I thought we had to be inspected and get a licence first, We can't do that before Saturday," said Den.

"I gather the customer's brother is on the licensing committee, and I think a friend of Mr Romanov's is too," said Mrs Lepanto.

Iska nodded. "Yes, with both of them batting for us it ought to be a given, surely."

"Can Cook manage on her own now? Come sit down, Maria," said Lucrezia.

"Yes, the coffee's ready on the sideboard, and the chocolates."

Den brought a chair to the end of the table and helped the house-keeper into it, then brought over the coffee pots and chocolates.

"It's amazingly short notice," said Lucrezia, cutting herself a slice of Pavlova. "Do we know what's behind that?" She looked around the table.

"Yes," said Maria Lepanto. "There's been a landslide. You may remember we had some nasty rainstorms last Friday. Apparently they caused an overhang to fall and the Castle Hotel has been damaged. They've had to cancel all their functions."

"What made them think of us?" asked Cesare. "We've never done any catering before - or have we?" He looked around the table.

"Well, we *have* done a barbecue now and then for the hang-gliding club, and we do the odd garden party for the disabled kiddies. You don't mind, do you?" asked Den.

"No, glad to hear it," said Cesare. "What did you do with the wolves?"

"Well, they weren't so much trouble in the past. Played with the kiddies – though we had to keep an eye on them, of course. It's just this rogue one that's growing too big for its boots. I think we ought to find it a place in a zoo. We can't afford to let it breed. Its cubs might be even fiercer."

"Maybe the vet would agree to it being spayed, and keep it in his kennels over the weekend while it recovers," said Lucrezia.

"Great idea," said Iska. "If not, we'll shut it in the stables. The others should be fine without it setting them a bad example."

"How many of there arum?" asked Den. "A barbecue's

103

easy enough whatever the numbers, but if they want a sit down do, how can we cope?"

"Only forty-four," said Mrs Lepanto. "They were hoping for something elegant and formal. That's why they'd like to come here. We'd find it a problem to cope with a larger number straight away, with no experience. Forty-four seems ideal, from the kitchen's point of view, enough to stretch us, but not to overwhelm us."

"You sound quite keen to give it a try. Do you think the staff will be happy?"

"Well, My Lady, I did take the liberty of asking around, and they seem quite excited. It can seem awfully quiet in this big house, and when we've just finished shining it up it looks so wonderful we want to hear somebody say so."

"I know exactly what you mean," said Justin. "It seems to be waiting to be admired, and it should be."

"We gave them a notice to put up in the delis when they made the deliveries this afternoon. I hope we did the right thing. Mr Fermi told me you had decided you wanted to do weddings and the problems at the Castle Hotel seemed just the opportunity we needed. I thought the notices might give us some idea if we might be a popular choice. The shopkeepers were very interested. One phoned these people there and then. They'd like to come and discuss it right away,"

Sitting next to Cesare, Justin felt his silent sigh. Yes, imagine knowing you were about to lose all this. But he had the means to save it. Why did he have such scruples? He was a warrior, admitted to killing hundreds, thousands in battle, and last week he had cheerfully dispatched another three, apparently in cold blood. Why had he invented this constraint for himself? A warrior who would only kill to save lives, or fund the hospitals, but not to line his own pockets. Yes, it did make sense. Less reprehensible than a simple lust for killing.

"So, do we know what these people would like?" asked Lucrezia. "Elegant and formal. That must mean staff waiting at table, in formal dress, I suppose. That means you, Den, in full butler gear - "

"Mr Fermi, if you please," smirked Den. "And Prince Alexander Romanov, you'll have to swallow your pride and be Romanov, or maybe just Iskander, the under-footman - under me."

"You'll see the underside of my foot in a minute, Fermi," grinned Iska.

Den screwed up his napkin and flung it at Iska who lobbed it back.

"Hey, watch out for the glasses! We'll need them on Saturday," Lucrezia exclaimed. "If you want a fight just go outside. Actually, if they're expecting the full country house thing, you'll be Fermi to me, Den, and Mr Fermi to you, Iskander."

"And what do we call the notorious Cesare Borgia, Marquis of Monrosso?" asked Iska. "Remember, princes outrank lowly marquises."

"Do they?" smirked Cesare. "Come outside."

"I wish I knew what it was I was supposed to have put in the wine," said Lucrezia. "I should spike your drinks every night. See if it improved your table manners."

"Valium?" suggested Justin. "That might help."

"Couldn't have been Valium. Hadn't been invented in the fifteen hundreds."

"What about tables?" asked Cesare. "How many does the Great Table seat?"

"Twenty, comfortably. And then there are the two big extensions. And the Morning Room table. We could make a three pronged 'E' shape, so they would feel more cosy," said the housekeeper.

"The dining room is big enough for that," said Lucrezia.

"Do they want to dance?" asked Justin.

"Well, forty-four would look lost in the ballroom, wouldn't they. A bit sad. And they wouldn't all want to, or be able to."

"What about the entrance hall," said Justin. "It's so gorgeous and would be fine for a small group of dancers."

"Actually I don't think people usually dance after lunch. They usually just sit about chatting and then go home mid-afternoon," said the housekeeper, "so that makes it even easier for us, doesn't it?"

"Tours of the House?" suggested Justin.

"Especially Iska's bedroom," said Den. "You'd have to move that train set and all those model drones – or we could tell them you're the teenage son and heir."

"And you'd have to dismantle all those experiments. Looks like Frankenstein's lab," said Iska.

"What? You? You're not the real Enrico Fermi, are you?" gasped Justin, "making another atom bomb - "

"You mean the one who died in 1954? Do I look like a zombie?"

"Well, I think we've done all we can tonight," Lucrezia broke in hurriedly. 'Mrs Lepanto has had a long day. She's entitled to some rest."

The housekeeper took the hint, rose, gathered up the pudding plates and bade them goodnight.

"Sorry," said Justin shamefacedly. "Not used to being a you-know-what. Can anyone else overhear us now?"

"Well, we do try to make a bit of effort not to give the game away - "

"You do what?" scoffed Cesare. "You almost shout it from the rooftops. Maria Lepanto must be suspicious – and with pretty good reason."

"Mrs Lepanto's a very wise woman," said Den. "Knows

which side her bread's buttered. Important job. Treated with respect. Gorgeous place to live. Some of the staff are third generation, have been here forever. If they think anything odd's going on they know it won't harm them and they won't risk losing their homes by outing us, will they? They get free homes, free food, free power. They've got too much to lose."

Justin glanced at Cesare. He had no need to be a psychologist to read that agonised expression.

And if the staff did lose their homes and livelihoods, would they still keep their suspicions to themselves?

What would the newspapers pay for stories like theirs?

CHAPTER 11

GIVE MY LOVE TO LEKRISHTA

Everything happened at the Dairy. It was the antithesis of the Great House. The Great House sat serenely on its hillside above the lake, as beautiful, and almost as quiet, as a mausoleum. 'The Dairy' was a collective noun for a factory and a miniature village. Half a dozen cottages, straight out of picture books, dotted a ridge just high enough to have a view of the lake through the windbreak of trees on the opposite ridge. In the sheltered hollow in between was a miniature village green with three kiddies playing on swings and a merry-go-round.

The dairy itself, an elegant two storey classical stone building, formed a barrier at one end of the hollow, and behind it were all the less fragrant edifices, the stables, the mistals (cowsheds), the machine stores and the workshops, all in fine classical style buildings arranged in a square.

Justin stopped the jeep in the middle of the courtyard and looked around for the workshop. The doors were wide open and an old man was whistling as he dragged four smart new bee hives into a line at the entrance.

"Mr Chase?" he called. "Here you are, Sir, all done and dusted."

"Hmm, very nice. Have you made them yourself?"

"Well, can't sit around all day doing nothing, now can I? Just not used to it. Will they do, do you think?"

"They look first rate to me. Are you Roberto? I'm sorry I don't know your surname."

"I am. I've always been Roberto. Must have a surname but blessed if I can remember it," he laughed.

"Didn't I hear you had retired?" asked Justin.

"Well, in a manner of speaking. It's the scrheumatics — got the better of me. Well, I'm 75, so only to be expected, I suppose. Shoving great big cattle and that mad donkey around just got a bit too much, especially when it's freezing cold and pouring down."

"You worked till you were seventy five? Crikey Moses!" exclaimed Justin. "I heard your daughter wanted you to go live with her. Sounds like a nice lady."

"Yes, fine lass, really, but who'd swap all this for a flat in a city?"

"Yes, you've got a point," sighed Justin, thinking of his penthouse in Rome, once his pride and joy. "So, you still live in your cottage?"

"Well, it was only fair to let the new man have it. I don't need three bedrooms just for me. Mr Romanov's put me a fancy little chair lift up the stairs above the dairy. Lovely and cosy up there, and the view's great. Nice change."

"Are you all by yourself up there? It looks a big space."

"No. The two young lads have got rooms up there too, and there's a few big rooms for the lads and lassies who come in to help for the odd week or two when we're busy. Plenty of company." He took hold of one of the hives and Justin hurried forward to help load them into the jeep. "Mr Romanov thought to put them up by the Refuge."

"Yes, that's where he wants me to take them. The manuka bushes up there are covered with bees, so it seems a good idea. Do you suppose any will move in, or will we have to buy a swarm?"

"They say if you smear some honey inside they'll go in to get it and maybe decide to stay. It's a bit late in the year for swarming, but you never know."

"Well, sounds worth a try," said Justin.

Justin stopped the Jeep in the kitchen courtyard and

went in to cadge some honey from the cook. He had to winkle her out from an animated discussion as she waited with Lucrezia, Den and the housekeeper for the customers to arrive to discuss the golden wedding party.

When he climbed back into the jeep he found Cesare sitting there.

"Sh!" Cesare whispered. "Let's go."

He made little attempt to sustain a conversation as they bumped and lurched up the grassy slopes towards the Refuge. Once there he leapt out, grabbed a holdall and strode into the cottage. Intrigued, Justin watched from the doorway as he pulled out slacks and T shirts and two pairs of deck shoes and secreted them in cupboards.

"Well," he explained as he walked back outside, "Can't keep pulling that one about a Roman fancy dress party."

"When you take the quick way home?" suggested Justin.

"Exactly," said Cesare evenly.

A cold shudder swept over Justin. How could the man take his probable impending death so calmly?

"Let's get these bee hives unloaded." said Cesare,"

Justin breathed a sigh of relief. He was only playing boy scout, being prepared, that was all. Yes, but - if he did let his scruples win, it seemed highly likely he would lose Monrosso. What a choice!

What would I do, he wondered. Well, as a fully paid up very mortal coward, probably copy the estate manager and cop out - and spend the rest of my life squirming, agonising about all those good-hearted staff, out on their ears, and, even worse, those little kiddies in bomb vests out in Raqqa. While I swan around inventing 'daring' computer games he goes out there and lays down his life - again and again and again. Some men are made of different stuff!

"How did you get started – in this – removing the terrorists stuff?   Do the Americans sort of advertise - ?"

"Got kidnapped, about six months ago, along with an American and a teenager. Tried to put it to good use, talked to the brutes, tried to understand what made them tick, but there was no way any right-minded person could find any common ground. They had no humanitarian agenda at all, no desire at all to make the world a better place. Seemed to want everybody dead, or frightened out of their wits. Thought the next world would be designed exactly right for them. I could have told them life after death is exactly the same all over again, or for them, just oblivion, but they wouldn't want to believe it, would they? They were just raving psychopaths, killing for the joy of it, got all flushed and bright-eyed about twisting the knife, hearing the victims scream. No hope of talking them into anything rational - like trying to tame a rabid dog."

"So, what happened? Did you get ransomed?"

"Ransomed? Who'd ransom me? No, so they started torturing the kid on video, chopped his hand off, said his head was next if they didn't get a huge ransom. The poor kid was bleeding to death - and of course I tried to help him, so they started on me. At first I thought, oh good, I'll soon be taking the quick way out of this Hell – but then, what happens to the other two? We'd nothing to lose. I yelled, 'Let's get out of here!' and karate-chopped the nearest one. Then the American joined in. It was a shock to see how useful he was. So, in short, we left a cellar full of dead terrorists and got the hell out.

"The American came to see me a few days later. Turned out he was CIA, had been a Seal, so no wonder. Anyway, he told me there had been bounties on some of the brutes we'd chopped and he'd told his controllers I should get my fair share. I said I don't murder for money, but he said wouldn't the money be useful for our field hospital. And you bet it was! We were desperately short of everything. Then everybody kept asking, 'Can't you get us some more money for this or that?'

"So many casualties, so much suffering, and all done deliberately by those slavering psychopathic loonies. I thought if I sent them off towards the Heaven they wanted so badly, then the rest of humanity might have a chance of improving the only world we're ever likely to get. But you're right. They're like the Hydra: cut off one head and nine grow in its place. I'm just making martyrs out of homicidal maniacs, aren't I?"

"Maybe that was just a lazy coward speaking. You have to shoot a rabid dog, don't you? Eventually you've shot all the rabid ones. All the rest are sane and normal. There were millions of buffalo in the USA once. The colonists shot them right out of existence, didn't they? And there are no wild wolves left at all in England. It's not impossible to get rid of noxious creatures, is it?"

"Ethnic cleansing? Gas chambers?"

Justin admitted defeat with the appropriate Italian gesture. Yes, it was a far too tricky question. Easy to argue from a comfortable armchair. Hope to goodness he'd never have to make such choices himself.

That helicopter was coming straight towards them. Cesare beckoned to Justin, strode over to the porch of the Refuge and picked up a rucksack.

"Give my love to Lekrishta," he shouted over the noise, "but not until she asks what's happened to me."

"Lekrishta?" asked Justin.

"Krish. Lekrishta's her real name, Mrushan name. Enjoy the rest of your holiday, and don't let us work you to death."

Feeling rather sick, Justin watched until the chopper seemed to evaporate into the hot blue sky, then he turned to look at Monrosso. A view to die for?

Something told him that, desirable as it undoubtedly was, it was not just Monrosso that had swayed his decision. It was the fate of the people who lived there, and those kiddies in bomb vests a long way away in Raqqa. Alas, poor Raqqa!

He smeared the whole jar of honey inside the four hives and went into the Refuge to wash his sticky hands.

How long, he wondered, before Cesare was back here retrieving his clothes and trimming his Methuselah beard? Would he remember, in the turmoil of death, to think of the Refuge here at Monrosso? Or might he wake up under the rubble in the Borgia House in Rome again, or behind those dustbins in Peshawar? Or might he be lying helpless in some hospital, somewhere in this big wide world, desperate to die while they struggled to keep him alive?

Justin feared he'd had his last good night's sleep for quite some time.

CHAPTER 12

PREPARATIONS

Yes, it appeared that lunch was to be served in the Morning Room. The table was set for five, with large bowls of fruits and salads down the middle. But would anyone turn up? He could see Den and Iska and a couple of farm hands struggling with a huge roll of wire in the centre of the lawn, showing no sign of stopping whatever they were up to. And one of the five seats would definitely be empty, unless Cesare had thought the better of it and asked the chopper to turn around and bring him home.

His heart sank at the thought of having to break the news to Lucrezia. But maybe she would let him comfort her with a few hugs, like last time. He could only hope.

Den looked at his watch and said something urgent to Iska. Iska sent the farm hands off in the direction of the Dairy, and the two men hurried towards the House. It was five minutes to one when they strode into the room, looked at their watches and visibly relaxed.

"What's the hurry?" asked Justin.

"She who must be obeyed said one o'clock," said Den.

"Do you always do exactly as she says?" exclaimed Justin.

"Well, what's the point of electing a queen and then disobeying her?" asked Iska. "We tried the Athenian system in Mrusha, debating everything ad nauseam, but the rows, the fights – well, it was mayhem. Nothing got done, while we argued and fought about every stupid little thing. In the end Krish lost patience, jumped up on a rock and yelled at us to shut up and listen to her. It was amazing. Everybody went quiet. Everybody listened to

114

her. Things got sorted in no time, without any fuss. We had a big debate and voted to make her our queen. Everybody vowed to obey her. So we did. You can't imagine how much time and trouble it saved us. No need to justify every little thing we did any more; just say, 'Lekrishta's instructions; Lekrishta approved it.' Le-ta means the queen, so she's Queen Krish."

"Is there a king of Mrusha as well?" asked Justin.

"Chez," said Den. "He's by far the toughest and the bravest, and she's by far the cleverest of us all, so we were pretty pleased with our rulers. 'Ra' is Mrushan for the sun, the source of all life and power, so he is ChezRa. He protects us, leads us into battle, if need be. We know we can depend on him absolutely, so he's an ideal king."

"Cesare Borgia has a rather iffy reputation, doesn't he?" Justin ventured cautiously.

"Huh!" scoffed Iska. "Pretty well every single one of the power brokers of the past could be done for murder, corruption and war crimes today. You can't judge people of a different era by the standards of today. His soldiers idolised him, and the folks he ruled thought he was great. But, in actual fact, Chez only took on his identity in 1503, when the real Cesare and his father, the Pope, both died of some horrible disease, or maybe poison. It deformed their faces so badly that he had a good excuse to wear a mask. He'd fought under Borgia so he had no trouble imitating the voice and the mannerisms. It turned out to be a bad idea. With Pope Alexander dead he was hounded by all their enemies, was butchered again and again, so he stuck it for only about five years."

"Didn't they dig him up in Spain last year?"

"They dug somebody up," grinned Den. "But Chez had helped bury the real Cesare in Rome, so draw your own conclusions."

"And Lucrezia Borgia?" asked Justin.

"Ah, well, that's a bit different. Pope Alexander used her as a sort of bribe to get other rulers on his side. He kept marrying her off, then getting rid of her husbands so he could marry her again to someone even more useful. She was only seventeen when he dragged her away from her first husband. He put her in a convent to keep her out of trouble while he bullied her husband into agreeing to an annulment on the grounds that she was still a virgin. But poor Lucrezia was pregnant. Having the baby killed her. Krish was a nun with the job of keeping an eye on her, so when the poor girl died, Krish thought she might as well seize the chance to take on her life. The 'nun' was buried, and Krish became a ruling duchess, did a very good job for twenty years. 'Died' at thirty nine after her eighth child."

As the clock above the colonnade struck one, Queen Krish walked in, looked around and smiled.

"Thank you for coming, Gentlemen," she said. "Please sit down. I hope this is not too inconvenient for you, but I feel, in the circumstances, we should try to meet twice a day to make sure we are all in the loop. Has anyone seen Chez? It's not like him to be late for food."

Justin cleared his throat. "Well, he sends his apologies - asked me to give you his love, Lucrezia - " he tailed off lamely. "Did you hear the helicopter about eleven?"

She put her hand to her face and shut her eyes.

"Well," said Den, "what else could we expect? Has he ever let us down?"

She glowered at him from behind her hand and smothered a sob.

"Here we are, My Lady, Gentlemen," said Mrs Lepanto, opening the door for Heidi and Judy, one carrying a big tureen of richly aromatic pasta and the other warm bread rolls to the sideboard.

Lucrezia fought to put on a smile. "Smells very appetising. I think we may make less of a mess if you serve it out for us, Mrs Lepanto. A small portion for me, if you please."

The maids, flushed and smiling, brought the steaming plates to the table and gave each of the diners a little curtsey. Then they carried the bread basket and the salad bowl to each in turn.Very charming, thought Justin.

"Please thank Cook for us, my dears," said Iska, as the three women headed for the door.

"I ordered him to tell them he wouldn't take on any more assignments," said Lucrezia, in a strangled voice.

"Was that your head speaking or your heart?" asked Den, quietly. "After all these aeons of time you still can't accept that he's a warrior king. His duty is to save his people, not please the woman he loves. You were wrong to force him to choose. It's not like you, Oh Wise One."

"Then I must resign," she said, staring at her plate.

"No point," said Iska."We won't accept your resignation. You're irreplaceable, so I guess we're stuck with you. After all, you're right almost every time, so we've no right to grumble. Cheer up. He'll soon be back."

"In the meantime," said Den, "we should do our bit, have something really worth while to show him when he gets back. The golden wedding people are keen to come here, and we've agreed a menu with them this morning, so we've got to work flat out to get things ready."

"Yes, but what about the licence and the inspection?" asked Iska.

"Well, if we don't get them fixed in time, we'll just have to do it for free, tell everyone they're our guests, that it's a private party, I suppose," said Lucrezia.

"How's that going to pay our debts?" demanded Justin.

"It would be an advert, spread the word around.

Encourage lots of other people to patronise us."

"They'd all cry off when they discovered we expected them to pay," said Den. "Anyway, we're going to get our licence, aren't we, so let's not waste time worrying."

"So, let's get down to business," said Lucrezia. "We've fixed the menu. Lunch is to be served at one thirty in the Dining Room. We have enough tables for forty four people. There are plenty of matching chairs spread around the House we can bring down. Crockery? That's a problem. We used to have summer tea parties outside and evening parties with dancing and wine and canapés but our biggest lunch or dinner party was rarely more than twenty, so we'll have to mix the plates and cutlery."

"They pinch the spoons, you know," said Den. "Especially tea spoons. Ours will go like hot cakes, with the lovely shape and the Monrosso crest. Why don't we buy some cheap stuff, then we won't care?"

"Four dozen of everything would cost quite a bit. Couldn't we hire them?" suggested Justin. "That hotel that's been damaged - maybe it's only the decorations, not the cutlery and crockery. Shall I go talk to them, see what they could lend us? Better not to buy anything until we're sure we're going to carry on catering and have some experience of what it's really like."

"Mm, well, that does sound like a good idea," said Lucrezia. "Yes, why don't you do that. Give you a chance to see more of the area. I'm sorry we're giving you such a ruined holiday."

"Not at all," said Justin. "It's quite exciting, being in on all this, and I'm glad to be some use."

"What about waiters?" asked Den. "In the grand old formal days we didn't use waitresses."

"The girls will be disappointed," said Iska.

"We could find them some reason to be out there, showing off their frilly aprons," said Den, "but we should have men actually waiting at table. What about you, Iska? And could the farm hands scrub up to scratch?"

"What about the waiters at the damaged hotel? They might be free and willing," said Justin. "You could let the farm hands watch and learn for next time."

"That would be an easy way out," said Den. "It would be a relief to know somebody could be relied on to do the right thing. The farm hands might be chuffed to have a chance to dress up in white tie and tails, and strut about taking selfies. But where do we hire the livery? There are no wedding hire shops around here."

"But we've lots of genuine livery up in the attic stores," said Lucrezia, "real historic clothes. It would be fun to get it out. Of course it might be full of moths, but perhaps nobody would notice. So, is that all about the actual meal? Right, now, where do we serve drinks, pre and post lunch?"

"I'd be disappointed if I didn't get a chance to have a good long look at the entrance hall," said Justin. "If you don't think they'd want to dance, could they have drinks there? Then the hosts could watch all their guests arriving."

"Lovely idea," said Lucrezia. "And if any of them thought it was a shame to go inside in such lovely weather they could take their drink back out onto the colonnade. We'd need a few small tables and chairs for anyone who needs to sit down."

"They could have their coffee and brandy there as well, or on the colonnade, admiring the view, like we did at breakfast the first day," said Justin, "before we realised anything was wrong."

"So, we need to locate a few small tables and chairs. Maria and I will go right through the House and get out everything promising," said Den.

"What drinks do we serve?" asked Justin.

"What's wrong with Monrosso Rosso and Bianco?" asked Den. "And our Marquis Reserve Brandy and Liqueurs and Aperitivos? Sell very well, so why not?"

"You've got a vineyard?" asked Justin.

"Of course. Don't say you don't like it. You've downed enough to float a boat," said Iska.

"It's gorgeous. Didn't realise you made it yourself." said Justin. "So, this party will be a good advert for your produce as well, won't it. Should we have a display somewhere of stuff for them to buy as souvenirs?"

"Nothing too obtrusive and blatantly commercial," said Lucrezia.

"Of course," said Justin.

"So, what else do we need to think about?" asked Lucrezia. "We said we could take anyone who asked on a tour of the best parts of the House. The girls would love to do that. What about outside? Some people might try to go exploring."

"Yes, well, that's what we're trying to work on now," said Iska. "We thought we ought to clear a bit of lawn next to the house so that people could maybe get to the kitchen garden without tripping over the animals or wading in cow-clap. And we need to mark out a car park area. Lucky the ground is so hard at the moment, and there's no more rain forecast till after Saturday. Maybe more than a dozen cars might arrive. We got some posts and wire delivered this morning and we're just trying to work out where to install it."

"Why don't we take our coffee out and sit in the colonnade, then we can see what we're talking about," said Lucrezia, as the girls brought in coffee and mints.

"What's going on?" asked Justin. There was a huddle of animals crowding around the reel of wire, cattle poking it with their horns and goats wriggling through the cattle's legs to get a better view.

"They're just interested in anything new. Got nothing else to do but eat, and they're not stupid," said Iska.

"What will they do on Saturday when they see all the cars and guests?" asked Justin.

"That's exactly what we're wondering," grinned Iska. "We've got to try and second guess them and be prepared. It's different for all those estates that have been open to the public for years. Their animals have lost interest in the visitors, but ours, now - who knows how they might react?"

"Somebody will have to man the garden doors or Rumpelstiltskin and his mates will chomp up all your veggies," smiled Justin.

CHAPTER 13

FANCY DRESS

"Excuse me, Mr Chase, are you busy? Could you spare a few moments to help us?"

"Yes, of course," said Justin. "How can I help you?"

In truth he was feeling rather spare. Straight after lunch Iska had headed off towards the Dairy, after rattling on about opening the main gates to let the animals have a good look at the cattle grids. Why on earth? Den had dragged off Lucrezia to help scour the house for small tables, since Mrs Lepanto and the servant girls had other fish to fry. Was he being invited to play the fish?

"If you wouldn't mind just coming upstairs with me, Sir."

Oho! he thought, scanning the trim, very middle-aged figure and fighting to suppress a grin. "The mind boggles! Lead on, Macduff."

She led him up the glamorous curving staircase, along the corridor, through a less heavily decorated door into a cleaners' room, then up a plain staircase to the top floor.

Goody good, he thought. This should be interesting. Servants' quarters. Blatant class distinction. No glamour and glitz up here.

It was a long corridor with a window at the end. Most of the room doors were open, with the sunlight pouring in. He stopped to look into the first room and gasped. It was a work of art. True, there were no gilded cornices, just simple elegant ones all painted white, but the walls were a riot of colour: country scenes in lovely pastel colours.

"This is so like the Marchesa's suite in Rome – or was," he said. "Such a shame that's all gone now."

"It's my sitting room. I chose it for the lovely art work. All the rooms up here are lovely. I gather the Masters long ago had a painter friend who couldn't get any work, so they took him in and asked him to make this floor beautiful. I gather the servants were allowed to ask him to paint anything nice they could think of, so every room is different. We can choose any room we like, as most are empty now. I could have a room on the floor below, like Mr Fermi, but it would be lonely for the girls and Cook to be up here on their own, wouldn't it?"

"How many people normally live in the House?" he asked.

"Well, for the last twelve years or so, only Mr Fermi and Mr Romanov on the floor below and the four of us up here. Occasionally there are guests, usually quite lively ones, so that's a bit of fun. It can be lonely and scary in the winter when it's dark and the wolves are howling. That means there's intruders prowling around."

"What do you do about intruders? Will the police come right out here?"

"We don't call them. We just leave things to the wolves." She laughed and bustled off along the corridor.

Oh dear! thought Justin. Woe betide any clueless visitor who went out and came back in the dark alone.

"Have you found any chewed-up remains of burglars about the place?"

"Oh, no. The wolves love crunching bones. They don't leave a thing. Their jaws are twice as strong as huskies and alsatians."

"Don't they scare you at all?"

"That young one did, scared everybody a little, but we have the dart guns. I gather the Marquis has taught it better manners now. We were sure he would. Here we are, Mr Chase. What do you think of these, then?"

"Crikey Moses! What amazing fancy dress! What are you thinking of doing with all this?"

"Mr Fermi thinks the farm lads might like to learn to be footmen if we decide to carry on catering, just to earn a little overtime and experience something different. They'll need to watch the first time to see how it's done, so we thought we might as well make them look special if they're going to be standing around doing nothing."

"You're surely not thinking of making them wear this? I thought footmen just wore white tie and tails."

"Nowadays they do, but this is what they wore when the House was new in 1745. It was designed to match the decorations. The Marquises at that time loved yellow. The customers asked for something elegant and formal. What could be more elegant and formal than this?"

"I should think it would exceed their wildest dreams," he laughed. "Do you really think the farm hands would stand for this?"

"Well, we want to see if it's still wearable after more than two hundred and seventy years. Would you please try it on for us. We don't want to ask the farm hands to come up here to try it, then find it's not wearable."

Justin tried hard to think of a reason to refuse, but extreme embarrassment didn't seem good enough. Very reluctantly he took the bright yellow knee length breeches and the yellow, gold braid-encrusted waistcoat into the next room and struggled into them. The women all whooped when he rejoined them – and no wonder, he thought.

"Now, the jacket," said the housekeeper. She passed him a black tail coat with gold braid and brass buttons. "Now, what do you think?"

Justin turned to look in the mirror with some trepidation. Wow! Who on earth was that? Some fabulous film star in a costume drama. It was impossible not to grin at his reflection. For the first time in his life he knew he looked stunning.

"We'll need some new white stockings and white ties. The old ones are grubby and dropping to bits. Could I try this bow tie on you?"

"Yes, that's better. The neckline looked wrong before," said the glamorous film star. "What about footwear?"

"We've found some grubby white shoes. Maybe we can whitewash them. Black would match the jacket, so they might have some of their own that would do. Could you walk around and sit down a few times to see if anything might disintegrate?"

"It's surprisingly soft and cool to wear," he said. "I thought it would be all stiff and hot and heavy."

"There's another set for winter. This is summer wear."

"I didn't imagine the aristocracy cared a fig for their servants' comfort."

"I expect there are bad employers in every walk of life, but, you know, it's not wise to make your servants hate you. They have so many opportunities to make you pay for it. Well, do you think this idea is worth trying? It should get everybody talking, shouldn't it? Make us famous."

"Look pretty amazing on people's selfies. How did they wear their hair? I've a feeling they had huge white wigs."

"Well, most butlers made them powder their own hair, which they absolutely hated. It ruined their hair after a while. Here they wore white wigs or, if they had nice hair, they just tied it back with a bow. Footmen were chosen for their good looks. Lucky our farm hands fit the bill."

"So," said Justin, "that rules me out, doesn't it? How many outfits do you have?"

"Eight of everything in four different styles, but these are the most glamorous, we think. The one you are wearing seems a good fit."

"So it does," said Justin, stealing a last glance in the

mirror before handing over the flattering black tailcoat with its yellow braid-encrusted cuffs. Those flunkeys must have outshone their masters. How odd!

"What do you think of the bow tie, Mr Chase?"

"Very nice. Much bigger than usual, but it looks right."

"It's the same size as the originals, and we'd never find any that size for sale these days, would we, so we've made it out of paper – kitchen roll."

"Very clever!" he laughed, as she took out the pin and removed the neckband. "Well, have you any other fun jobs for me, or am I surplus to requirements now?"

"You could tell the Marchesa all the livery is still wearable, so we could do a dress rehearsal for the farm lads whenever Mr Romanov can spare them."

Lucrezia was still fully occupied, this time in the kitchen with Cook, Den and a stranger. The Inspector? Nothing to be gained by joining them: he knew absolutely nothing about running restaurants. Better make himself scarce – or try to deliver the message to Iskander himself. It should be a nice walk down to the Dairy. Why not go via the gates and see if the animals really were inspecting the cattle grids.

They certainly were! There was an assorted gaggle just inside the wide open gates, jostling each other and poking the nearest slats with their hooves. The metal slats were loose and twisted around when pressed, which the animals seemed to find alarming. Justin wormed his way to the front and tried the grid for himself. Easy peasy. Big flat human feet were like snow shoes, bridged the gaps between the slats, so he had no difficulty whatever in walking across the grid and out of the gates. They watched him forlornly as he glanced each way along the road. A truck roared past, then a bus came hurtling from the opposite direction, stirring up a whirlwind of dust. He waved at the beasts and shouted,

"No, don't even think about it. You wouldn't last two minutes out in this nasty world."

Did they somehow understand? Well, they seemed more interested in him now, following him like dogs as he set off for the Dairy. He felt a shove against his back and turned, to find a huge shaggy orange face only inches away from his own. Crikey Moses! He'd thought only the wolves might eat him. Surely Highland Cattle weren't carnivorous. He aimed a blow at the huge face, somewhere between the big scary horns, and the cow stopped in its tracks and stared at him. Was that a bit unkind? He stretched out his hand and tentatively rubbed its face. Out came a huge slimy tongue to lick his hand. Yeuch! Well, better than being butted in the back. He rubbed its face again, then felt another shove. Another cow seemed to want its share of attention as well. And then something butted him behind the knees, so his legs almost folded. The goats were joining in. Get walking, you twat, before the whole lot of them start butting and trampling you to death.

By the time he reached the Dairy he was jogging – and so were some of the beasts.

"Shooo!" yelled one of the farm lads, "You've been milked already. Buzz off till tea time." Waving his arms he shooed them out of the yard.

So, that's how you do it, thought Justin. But would they take the same notice of me?

The farm lads reacted exactly as Justin had done when he described the livery he had tried on - obviously intrigued but embarrassed.

"You have to be good-looking to be a footman so that probably rules me out, but Mrs Lepanto and the girls think you two should qualify."

They looked at each other and pulled shy flattered faces.

"When Mr Romanov can spare you the ladies would like you to go up to the House and try it on, see what you think. I gather there'd be overtime pay, and you'd be right in there with the guests, see and hear everything. Better than watching Downton Abbey on TV."

"I'd be a bit nervous, like," said one of them. "No idea what to do."

"You'll just have to stand there watching and looking gorgeous this first time. Mr Fermi will be in charge, and I expect Mr Romanov and maybe myself will actually do the waiting on. We're trying to get a couple of real experienced waiters from that hotel that's out of action. Then we can all learn from them. It's only a lunch party for forty-four retired people, so it won't last long. Should be a bit of fun. A perfect chance to practise. Where's Mr Romanov?"

"He's with Fabio and Giorgio trying to work out where to put the fences – for the car park and to stop the animals bothering the visitors. They're bound to be interested. Nothing else to do but eat, have they?

CHAPTER 14

ONE MAN'S DISASTER IS ANOTHER'S OPPORTUNITY

"Oh what a mess! You must be feeling pretty low," said Justin. The hotel had certainly taken a nasty hit. It was loud with the hiss of hoses and the scrape of shovels. Filthy curtains hanging by their last few hooks trailed into the mud and there was a large heap of muddy furniture in the foyer. There seemed no end in sight to the clean-up.

"I'd be feeling even lower if the place still belonged to me," said the manager. "Thank God I sold it last year! Heaven knows how I could have got the place back together again. Selling out to the chain was the best thing I ever did. They're taking it in their stride. You'd think they coped with stuff like this every day of the week."

"I hope you're well covered by insurance."

"Yes, it's amazing they didn't put an excess on that overhang, but it had been like that for ever. Nobody expected it to fall. So, we're to have a total refit in their house style. I suppose a few old fogeys will complain, but we're all delighted to see the back of the old equipment. I couldn't afford to replace it. Now the finance isn't my problem any more. I can just enjoy running the place. So,the Monrossos are branching out into catering. That's a change. Are they letting rooms as well?"

"It's all very experimental at the moment. They've no experience, so they want to take it slowly, just the odd small special occasion, like this Golden Wedding you passed on to them. I suggested to them that you might be prepared to give them a helping hand, maybe rent them stuff like cutlery and bar tables. Do you think any of your waiters might moonlight for them on Saturday lunchtime?"

"It's possible. I'll give you their numbers and you can fix it direct with them. Furniture? Well, Head Office wants rid of all this lot. It's too old-fashioned for their image. It's covered by insurance, so they're writing it off. If you think it's any use you can have it. You're just in time. I've got a truck coming to take it away in half an hour. We can't clean up the foyer with all that junk piled up there. The insurers say it's worthless, so anything you can't use just scrap. How about you paying for the truck and re-routing it to Monrosso?"

Justin led the truck through the open gates of Monrosso, hoping he was the returning hero, not the purveyor of a load of nuisance trash. He had agonised all the way about where to have it unloaded, as he hadn't been able to contact anyone by mobile. Dumping it on the lawn next to the entrance to the kitchen courtyard seemed best. He could surely get a hose connected there and swill the mud off. Then it should dry well in this hot weather and they could pick out anything useful to carry into the House. Hopefully they could move the useless rubbish away with the machinery he'd seen down at the Dairy. Would two days be enough to sort it all out? Two days was all they had.

There seemed to be nobody about in the Great House, so he had to go ahead and have the truck unloaded and pay off the driver. Crikey Moses, what a load of disgusting filthy rubbish!

He was quite hot under the collar by the time he found enough hoses to reach the tap and the filthy heap, but at last his clean-up was under way. It was a dispiriting job. Everything looked so tired and tatty. Maybe it could be painted up. He slogged on, lining up wrought aluminium bar tables dripping forlornly onto the grass. Some were too wrecked to stand, so he started a heap of scrap. The chairs were a similar Art Nouveaux design, so he arranged them around the tables like a café.

The animals, predictably, began to arrive to supervise proceedings. He tried a waft from the hose, but they obviously enjoyed that and pressed closer, so he had to try waving his arms and going, "Shooo!" Thank God they took the hint! One by one they sat down to watch, chewing amiably.

At last Cook appeared at her door, attracted by the sound of running water. "Mr Chase, wherever did you get all those? That's exactly what we need, something we can put outside and leave there."

"Outside where?" he asked warily. Was she telling him they were only fit for the municipal dump?

"On the colonnade. Don't you think so? And at the bottom of the steps below the colonnade.   You could all have your drinks there, couldn't you, in the fresh air. One or two sun umbrellas would look very welcoming for visitors, wouldn't they? They've got nice marble tops."

"Have they? Well, well, so they have. You think they will do, then?"

"Perfect. Just what we need. You look as if you need a mug of tea. I'll go make you one and you can sit at your new table drinking it. Oh, these look like napkins. What a mess! I'll put them straight in to wash." She pulled an armful of grubby white somethings out of a soggy cardboard box and headed back to the kitchen.

She was back with two mugs of tea in no time, put them down and dived into the remaining pile of unsorted mess with enthusiasm, homing in on the small things while he finished extracting the large ones.

"Cutlery. Oh, you lovely man! Lots and lots of cutlery. It's quite nice, look. No more nightmares about them pinching my lovely crested spoons. Are there any plates?

131

We really need some big don't-matter-plates. Oh, all broken, oh! Now, here's some more, and these are better. And I don't think they're broken either."

Justin helped her carry them to the kitchen door, then refused to go inside dripping muddy water.

"You're a nice man," she said. "I'll ring the Dairy and see if they have anybody free to take your tables round to the front, once they've done the milking."

Milking? What about all the alpacas and goats and cows chewing their cud up here? What cows? They'd simply disappeared. Didn't they have to round them up with dogs, or horses or something? Apparently not on this farm, where the animals seemed a law unto themselves. And where were the wolves?

"Gone hunting and exploring, up in the mountains, I expect," said Cook. "They're wild animals, you know."

"Do you know what's happened to the Marchesa?"

"She went down to the Lodge to work on the accounts. Mr Fermi has been helping Mrs Lepanto get the young lads fitted out in the old livery. They came down to see where they will have to stand in the dining room and he took them through what the meal will be like. I had to pretend to be a guest and look at them now and then, so they knew to come and ask if they could help me. I'm going to put extra salt cellars and cutlery on the sideboards so they can get those, if need be. If it's anything else they are to say they will get you or Mr Fermi or Mr Romanov to deal with it."

"So, does that mean I'm to be a waiter? That should be fun, but I don't expect they'll let me wear the livery."

"Mrs Lepanto and the girls say you looked very dashing and they hope you'll be wearing it on Saturday."

Life suddenly looked a little brighter.

The next two days passed with everyone on the verge of panic, rushing around and wailing that nothing was as it should be and Saturday would be chaos.

But, of course, it wasn't. At eleven thirty on Saturday morning the whole cast sat on the colonnade, eating sandwiches on paper plates, drinking orange juice and coffee from paper cups and surveying the scene. Parasols had been brought up by the families in the cottages, luckily matching yellow ones, and the scrap tables had dried into a semblance of historic opulence. Somebody had even come up with yellow seat cushions.

The fourteen people sitting there looked like actors on a film set: Cook in her usual whites, the Housekeeper and the maids, plus two of the wives, all in elegant black with white trimmings, and the seven men in yellow and black Eighteenth Century splendour. The two waiters from the Castle Hotel had been so envious of the livery that Mrs Lepanto had rushed them upstairs and kitted them out too. They were taking selfies with everyone else in sight, and, of course, with Lucrezia, looking dazzling in a pale yellow dress and heels. It was a moment to cherish, whatever events may follow.

"Good luck, everybody," said Lucrezia. "You've all been absolutely wonderful preparing for this afternoon. Enjoy the party. Let's give the guests a very good time. Show them what Monrosso can do. When it's over let's all meet out here for a celebration."

Cook and Mrs Lepanto gathered up the paper crockery, put it into a bag and took it out to the dustbin.

As the first car came slowly up the drive the staff formed a fan outside on the steps, while the two waiters poured out glasses of Monrosso Prosecco on a table hidden by the sweeping curve of the beautiful staircase in the Grand Entrance Hall.

Lucrezia stood alone at the top of the steps, every inch The Queen.

CHAPTER 15

IT ALL WENT WITH A BANG

As the car scrunched to a halt, Iska walked down the steps to open the doors.

Mrs Rubinstein struggled out and smoothed her new dress. Then she caught sight of the resplendent fan of staff decorating the grand colonnade.

"Sweet Jimminy Christmas, look at that!"she whispered to her husband. "Wait till your brother sees all this!"

"May we park your car, Sir?" Iska beckoned one of the liveried lads and handed him their key.

The kid's eyes glowed. This big new Merc was way beyond his wildest dreams. Iska took a sharp breath and rolled his eyes. Well, the kids had been practising parking for hours, once they'd marked out the car park with a wire fence. They'd had the Jeep, a couple of small Fiats, an old Alfa Romeo Spider and a battered old Merc automatic to practise on, but they were not in the same league as this.

Lucrezia floated regally down a few steps to meet the hosts and led them up to the palatial entrance hall. The waiters from the Castle Hotel stood smirking self-consciously in their borrowed finery, balancing trays of Monrosso Prosecco.

"To you, dear Mrs Rubinstein, Mr Rubinstein," she said, handing them each a glass and raising one herself in a toast. "My husband, the Marquis, and I wish you a very happy Golden Wedding Anniversary. Such a long and happy marriage is too rare these days, and we feel privileged to welcome you to Monrosso to celebrate your very special day. If there is anything you need please don't hesitate to tell my staff and we will do our best to help you.   Ah, I think your guests are arriving.   Please excuse me while I go to welcome them."

"Well, will this do for you, honey? You can't say I never take you anywhere, can you?" grinned Mr Rubinstein.

Standing to attention just inside the open doors, Justin was ideally placed to see the expressions on the faces of the guests. The Great House was stunning, and today there would be plenty of people to appreciate it. He glowed with pride as if it were his own. It had to be saved, saved as a living home, the historic home of a Marquis and his lady, not as an empty, soulless museum.

In the kitchen Cook and two of the wives were bustling about, getting tiny savoury nibbles out of the oven and onto small silver trays.

"Calm down, ladies," warned Mrs Lepanto, as the parlour maids seized the trays and scampered towards the exit. "Dignity, if you please."

The parlour maids tripped through to the entrance hall, eager to see the guests – and their own overdressed colleagues.

"Did you see that frock? OMG!" screeched Judy, as she bounced back with her empty tray.

"And Gianni and Luigi, did you ever imagine they could scrub up like that? Wow!"

"And the Marchesa!   She's so beautiful."

"And the Marquis, What a shame he's not here! He's just - wow! Wonder what he would wear."

"Not the livery. That's just for servants."

"That's mad, isn't it? The servants get the best clothes. Weird!"

"Mr Fermi shouldn't be wearing it either, but I don't expect any of the guests will realise that. He couldn't resist joining in the fun," said Mrs Lepanto.

"Why shouldn't he?" asked Judy.

135

"Butlers are too important. I expect the Marquis would just wear a nice suit as it's lunch, not dinner, and, in any case, it's not his party. He and the Marchesa are not even invited. Monrosso is just a restaurant today. Really nobody should be in livery, as we're not serving aristocrats today, are we?"

"Well, then, why are we?"

"Well, obviously we want to make a big impression, get everybody talking. Get lots more customers. And it's fun, isn't it? You like seeing the men dressed up like film stars, don't you?"

"Why can't we have nice coloured clothes as well? Not fair, is it?"

"Why don't you suggest it to the Marchesa?"

"I will. Then maybe next time -"

"Time to put out the first course, ladies," called Cook.

Den walked solemnly over to the huge Chinese gong and hit it hard. "Ladies and Gentlemen, Lunch is served. Please follow me."

He led them past the farm lads, standing rigid with embarrassment along the corridor, to the entrance to the grand dining room.

The tables were works of art. Large pieces of Sevres porcelain in a striking shade of turquoise were arranged along the centres, holding artful arrangements of fruit and flowers. Side plates and place markers alternated between turquoise and deep pink.

Den directed the guests' towards their places, Iska, Justin and the waiters led them there, helped them into their monogrammed chairs and placed their napkins on their laps. Then they began filling the shimmering glasses with Monrosso Bianco, Rosso and water.

The farm boys followed the last guests in and stood to attention in front of the sideboard while Den took up station behind the hostess. He had perfected the art of appearing quite impassive while his eyes relentlessly searched out any tiny fault.

Waiting at table proved a revelation to Justin. He had dined at many of the world's best restaurants, and had long ago lost his fear of supercilious waiters. Nowadays he accepted their service without paying it any attention whatsoever, and so did this erstwhile Russian prince. Den had drilled the pair of them mercilessly each meal time the previous two days. He had trained them well, so they quickly formed a team with the two slick professionals. All four of them smirked with self-satisfaction at the attention their staggering appearance was generating. A welcome change from being treated like part of the furniture.

"I'm not sure they're enjoying this much," Iska muttered as they passed each other in the kitchen. "Very subdued, aren't they? Maybe this wasp livery is all a bit too OTT."

"A change to see Americans right out of their comfort zone," grinned Justin. "Enjoy it while it lasts."

The drink, of course, soon loosened everybody's tongues and doled out the Dutch courage.

"Tell me, young man, do you wear these amazing outfits every day?" Mrs Rubinstein demanded.

"Only in the summer, My Lady. In the winter we have the same livery in warmer fabrics."

"Well, dip me in chocolate – ! So every day you waltz around in this!"

"Normally, My lady, we reserve our livery for our Lord Marquis and his Lady, but they decreed we should extend this courtesy to you, Mr and Mrs Rubinstein, and your distinguished guests," purred Iska.

Nobody does smarm like the Russian aristocracy, thought Justin. Maybe I should ask him for lessons.

"And speaking of the Marquis's Lady, what's she up

to? Playing the secret restaurant critic?" said Iska, as they passed each other again.

"Don't follow you," said Justin, as he picked up plates of grilled rainbow trout straight from the lake, with Cook's special sauce.

"Don't tell me she's fooled you. Sitting near the bottom of the wing on your side. If she wanted to wear a wig you'd think she'd have picked a different colour. She just looks as if she's had a hair cut."

Justin took a different route back next time, down past the end of the wing. Yes, Iska was right. What a disguise! Lucrezia's red hair appeared to have been cut into an Anna Wintour bob with a long thick fringe, and she was wearing heavy-rimmed spectacles, a frumpy old lady flowered dress and very butch sandals. Trying hard to be a plain Jane and failing miserably, as usual.

When he'd finished delivering the main course, three kinds of meat, highland beef, alpaca and goat, all with delicious sauces, he again detoured on the way back to the kitchen. He offered to refill her glass and gave her a conspiratorial grin. "What are you up to, Lucrezia?"

"I'm sorry?" she said with a puzzled look. "No thank you. No, no more wine for me, thank you."

He smiled at her again, but she gave him a bewildered look. Well, maybe he was threatening to blow her cover, so he'd better leave her to her own devices.

Each time he looked in her direction, as he carried out the puddings, a delicious confection of freshly picked fruits, goat cheese and highland cream, he caught her looking at him with a puzzled frown. Very odd.

It was a relief to everyone when the speeches ended. The farm boys had long been shifting their weight from foot to foot and stifling yawns. But, of course, even bad things come to an end. The Rubinsteins elected to lead their guests back to the entrance hall and colonnade for coffee and liqueurs, and Justin seized his chance to waylay Lucrezia.

"Well," he said, "I hope you're feeling pleased with us all. So far so good."

"Yes, it was a very enjoyable lunch, and the service was impeccable."

She gave him a formal smile, then a dismissive glance, and turned to follow the other guests out of the room.

Hmm, thought Justin, I've obviously offended her. Anyway, shouldn't bother her with apologies now. Better wait till it's all over. He let her get a good distance ahead before following her towards the entrance hall.

What the ! - - ! Lucrezia was coming towards him, looking a million dollars in her simple pale yellow frock with her beautiful long hair flowing free. He watched her freeze, gasp, then walk hesitantly towards her alter ego. They both stopped dead in obvious consternation.

This was too good to miss. Justin sidled up to them cautiously, but they were far too wrapped up in each other to notice him.

"I'm sorry, I didn't catch your name," said Lucrezia cautiously. "Are you related to Mr and Mrs Rubinstein? I hope you're enjoying the party."

"I'm a friend of their grand-daughter. My name's Caroline Wilson. The waiter called me Lucrezia. Is that your name? I hope you won't be offended if I say that we two seem to look quite similar."

"That's exactly what I was thinking," said Lucrezia. "Amazingly similar. Let's look in the mirror. There's one just inside the kitchen, so the staff can check they look presentable."

They stared at their reflections, stood back to back, then held out their hands side by side. Lucrezia held a strand of her lovely hair against Caroline's short bob.

"This is amazing," she said. "You're Caroline - you're not Kerallyn are you?" she added in a whisper.

The other girl looked startled. "Please say that again."

"Kerallyn," whispered Lucrezia.

"Yes, I'm Kerallyn. And you? Lucrezia? Lekrishta! You're Krish – Lekrishta! You were darker skinned last time I saw you, but somehow I know it's you."

"And so were you, and we were so alike back then that people kept mixing us up, calling me Kerallyn."

"And they called me Krish. Oh it's so thrilling to see you again!"

They hugged each other ecstatically.

"This is too good to be true," said Kerallyn. "I've been alone so long. Every time I find one of us they soon die and disappear. I get so lonely."

"It's happened to me as well so many times, but there's five of us living here now, so there's a good chance we can help each other to stay together."

"There's four more of us? Where? Where?"

"Behind you. The earwig. Do you recognise Justin?"

"Justin? Hello, Justin. So thrilled to see you again." She held out her arms for a hug.

Yum yum! thought Justin. She felt so like Lucrezia, but far more user-friendly, tactile and uninhibited, not reserved like Lucrezia.

"The other three? Are they here? Can I meet them?"

"You've seen them already, waiting at table. You didn't recognise them?"

"Well, we were all dark-skinned in Mrusha, weren't we? There don't seem to be any brown people here today. Have they all lost their colour too?"

"Iska!" Lucrezia called him as he breezed into the kitchen and picked up two silver coffee pots.

140

"Yes, My lady," he called, then did an astonished double take, put down the coffee pots and hurried over.

"Meet Kerallyn," she laughed. "Isn't this wonderful?"

"Kerallyn? OMG! Come and be hugged," exclaimed Iska. "How you've changed! That's an understatement. But you've changed in exactly the same way as Krisha. How weird!"

"You two look quite different, though," said Kerallyn.

"Well, Justin insists he was never there in Mrusha, so he must be from some other village," said Lucrezia. "Denrico and ChezRa and Iskander still look like triplets."

"Coffee, coffee, come on with the coffee. Iska, you're slacking!" Den came storming into the kitchen, every bit the butler. "OMG! Who's this?"

"Kerallyn, and who are you?"

"Denrico. Kerallyn? OMG! What a lovely shock!" He gave her a big hug. "And now we are six. Things get better every day. Why did nobody tell me you were coming?"

"My name's on the list. You read it out, Caroline Wilson," she laughed.

The Castle Hotel waiters came breezing in. "They're asking for tours of the House and the garden, and we need more coffee."

"I'm getting in the way, spoiling the party. I'd better get back to my friend. Can we talk later?"

"Don't you dare slip off at the end. We don't want to lose you again," warned Lucrezia.

"Who do we ask about the tours?" the waiter urged.

"We're to do the House tours," called Heidi. "We clean the House, so we know it better than anybody. Mrs Lepanto's told us what to do and she's going to help."

"Can you manage the entrance hall and the tables outside on your own?" Den asked the Castle Hotel waiters. "There shouldn't be much to do till tea time. We'll all come back to help you with that. Right now the rest of us had better go outside to supervise the tours of the garden."

"When the animals see us all out there goodness knows what they'll do," said Iska. "They're bound to be interested."

"What about the car park?" asked Justin. "The animals might have -"

"They were very interested," said Iska. "I got the older farm hands to stand guard on that, don't worry. Now, better get the young lads to guard the garden. If the goats get through that fence they'll eat the lot."

It was Mrs Rubinstein's daughter who inadvertently started the fracas.

"Aren't they sweet?"

And, of course, they were. Purring beguilingly, the alpacas stretched out their woolly noses to be stroked, blinked their enormous innocent eyes and fluttered their irresistible long eyelashes. Oh, shame!

Two more guests joined the alpaca fan club, and more alpacas trotted up to the thin wire fence to claim their share of admiration.

Goats appeared from nowhere, always up for trouble, shoved and jostled the alpacas, begging for treats.

"Oh, no," groaned Iska. A wolf was worrying one of the orange pennants hanging from the wire fence. Everyone had forgotten the wolf pack. They'd been inexplicably missing for days. These fladry, hanging pennants, were supposed to frighten animals off crossing wire fences. Pull the other leg, thought Justin, who knew nothing at all about farming. As if a thin strand of wire with a few bits of bunting would keep a wolf from its dinner!

Unfortunately it seemed the townie might be right. Four more wolves joined in, fighting to get their teeth into the pennants, diving back and forth under the wire. Frightened of fladry? My foot! Oh dear!

The goats were getting too boisterous for the alpacas, whose purrs now turned to plaintive squeals. The wolves lost interest in the bunting. It was more fun to harry the squealing alpacas and the goats, rounding them up and making them run, teasing out the weakest that would be the easiest to kill.

A thunder of hooves and a herd of huge orange shaggy darlings came galloping up to the fragile strand of wire, shoving the smaller beasts out of the way and butting the wire with their great heavy heads.

The guests, drifting back from their tours, stood transfixed on the steps and the colonnade.

Iska turned and ran up the steps into the House,

What's the best policy when the farm manager has fled his post, Justin wondered. Try to get the guests inside the House before the animals break through the wire and charge into the House? Could they gallop up the steps? Could he shut the big front doors in time?

There was an almighty roaring bellow, and the Monster Donkey came charging into the fray, along with the llama, who wasn't going to miss an excuse for kicking something, just for fun.

The fence was writhing frantically. A cow had caught its great curling horns in the wire. It was tugging so hard the post was being pulled out of the ground. A couple of wolves came prancing, barking croakily and making lunges at the tangled up cow. Crikey Moses!

Bang!

The topmost branches of one of the giant cedars quivered and a gaggle of birds flew out screeching.

The animals froze, looked frantically around, then

143

scarpered, as fast as their assorted legs could carry them – even the wolves!

Everyone turned to look at Iska. He shouldered the shot gun and gave a smart salute, then slid it down to rest its butt on the ground. Then he bowed and smiled, first to the left, then to the right.

"Bravo!" shouted Denrico.

"Bravo!" shouted Justin and the farm lads.

"Bravo! Bravo!" shouted the shell-shocked guests. "What a show! Like Colonel Cody's Wild West Show. Buffalo Bill, Annie Get Your Gun. Fantastic!"

Iska bowed again, milking the applause like a pro.

Smarmy devil, thought Justin, pretending he'd staged the whole thing.

"How do you train all the animals?" asked one of the guests. "They all did their parts very well."

CHAPTER 16

ANCIENT ALCHEMY

"Iska, can I ask you a cheeky question?"

"No," grinned Iska. "Draw the line at cheeky questions."

"Pity,"said Justin."You said I'd a lot to learn about being immortal - "

"Well, if you're asking for advice, maybe. Go on then, what's the question?"

"You've all got interesting names. I don't mean your Mrushan names. Den's told me how Cesare and Lucrezia got their present names, but how about you? There's another man calling himself Prince Alexander Romanov, so won't he object?"

"I thought he might, but he never did and he died a few years ago. Anybody calling himself Romanov is a bit suspect. The Bolsheviks were convinced they'd shot the lot. I was there when they did it. We had to throw the bodies down a huge pit. There were so many of them they didn't bother to strip them. I guessed they'd shoot us too so we couldn't tell the tale, so I looked for a hiding place. I spotted a nice thicket, but they guessed my plan and – bang! bang!

"I felt sure I'd have to claw my way out of the pit next morning, but my body had other ideas. During the night it managed to leak out of the pit and hole up in the thicket. Pretty clever, except that I was completely starkers. My clothes were still in the pit. Fortunately it was summer and they'd left a spade behind. First thing I dug up was Alexander. What a coincidence! Alexander is another version of Iskander. He had papers in his pockets as well. So, off I tripped to Paris and joined the other titled refugees being wined and dined by high society. Nothing like a glamorous title if you need a free meal. What amused me was that the other so-called aristocrats pretended to recognise me. Obviously they were just as bogus as I was."

"So Cesare's the only genuine aristocrat around here."

"Genuine? Interesting concept. Well, he's personally held this title for more than three hundred years. Sounds a bit dubious to me," he grinned. "They are certainly not real Borgias, anyway."

"No, Den told me all about that, but he said Monrosso was Cesare's reward for his courage in driving a crack Swiss army out of this area. He's surely more genuine than any descendant who just inherits a title."

"Anyway, both titles have a nice swagger, don't you think, genuine or not."

"Well, then,Your Royal Highness, I gather you've kindly offered me the use of your state coach for the morning."

"Make sure you bring her back in one piece or it's sabres at dawn," he warned, handing over the key.

"Who, Kerallyn or this old Spider?" smirked Justin. He forced the ancient red Alfa Romeo into gear with a grinding rasp that made Iskander wince, then drove off quickly before the Prince could remonstrate.

It was the perfect vehicle, short of a motorbike, for a hot morning drive along the lake shore, and the Sunday traffic was fairly light all the way to Menaggio. There were so few cars waiting for the ferry he could park the glamorous open classic car where it could be seen to best effect. He was just in time - the boat was already drawing close.

He spotted her almost immediately. That red hair was like a beacon. His pulse began to quicken as he watched the boat drop its ramp and the cars begin to roll ashore. She was dragging a suitcase, so he hurried to take it from her, stood it up and held out his arms.

"Hello, Kerallyn, great to see you again."

"Hello, Justin. So thrilled to see you." She flung her arms around him and hugged him enthusiastically. "It was wonderful to meet you all yesterday – and such a shock. I was so excited I couldn't sleep. I haven't seen so many Mrushans for thousands of years. You're sure it won't cause problems having me to stay a few days?"

"Far from it. Everybody's longing to see you again. Let's put your bag in the car."

"Wow, I love your car!" she enthused.

"Would you mind if we took the time for a coffee before we set off to Monrosso? I need a break."

Fibber, he thought, but so what? These Mrushans had a pretty eccentric attitude to the truth, to put it mildly, so why not adopt the house style. What he really wanted was a little time alone with Kerallyn, before Iska or Den could pick her up. Well, yes, he'd little hope if one or both of those hunks took a shine to her, but if he didn't even try he'd certainly get nowhere, wouldn't he?

"There are some nice cafés in Menaggio, and we could walk around and stretch our legs first," she suggested.

"You've been here before?"

"I've been staying with Stacey's family over in Bellagio for a fortnight, so we've been over here a couple of times. It's only a few minutes across on the ferry."

"Did they mind your walking out on them?"

"No, no, they've gone. They took a taxi this morning to Milan for a few days' shopping before they fly back to Boston. They were going to give me a lift, then I was going to catch the train to Rome, so it didn't put them out at all."

"Why Rome? More sight-seeing?"

"No, I've got a six month sabbatical and I'm spending it all in Rome. I've got a secondment to the UN Food and

147

Agriculture Organisation. Just the chance I need. There's a little café right at the end of the prom, just a few tables looking over the water. Do you feel like walking so far?"

I'd love to walk you all the way back to Rome, he thought, but let's not gush like an idiot. "Why not?" he said.

"So, tell me," he said, trying to stir the irritating froth into his coffee, "what's a nice girl like you doing with the UN Food and Ag? I can't imagine you in muddy wellies."

"A white lab coat, you mean. My field's genetic research, diseases caused by genetic mutations. It's very exciting. New things are being discovered all the time. But I don't want to bore you with all that. Tell me about you. Lekrishta said you came from a different village. I didn't know there were any other villages besides Mrusha. Where was it?"

Justin sighed. Should he use his imagination, make up something that might capture her interest? The trouble was that he'd been brought up by a mother who was a stickler for the truth, and wouldn't tolerate bad language. As he couldn't stop himself from swearing she insisted he must use only innocuous curses. He'd had so little practice in lying he knew he'd make a mess of it. Then Kerallyn would be sure to despise him.

"The truth is, Kerallyn, I seem to be a bad case of memory loss. All I can remember is being a computer systems entrepreneur aged 29 from darkest Yorkshire – in the north of England. I can remember loads about those 29 years but that's the lot."

"So you're not really - " She put her hand over her mouth.

"They seem determined to believe I'm - immortal," he said quietly.

"Why?" she whispered.

148

"We crashed a helicopter, Cesare, Lucrezia and me." He kept his voice very low, even though there appeared to be nobody within earshot. "It exploded and burned out with us inside it. We woke up together next morning completely uninjured - a hundred kilometres away at Monrosso."

"So they can both confirm you are – you know?

"Mm. Iska wants to knock me off to see if I'm genuine. I'm trying not to let him get behind me," he grinned.

"That's not very friendly of him, is it? Well, I'll help you keep a lookout."

"That's very sweet of you. Now, tell me about you. You really are Mrushan, like the others, so you're a young lady of about twenty-five. Twelve thousand and twenty-five."

"You make me feel so young," she sang quietly, and they both laughed.

She reached up to adjust her sunglasses and the laughter froze in his throat.

"Is your husband in Rome or did you leave him behind in the States?"

She glanced down at her hand, then looked up at him with a sigh.

"His ashes are in the park, so I don't have to walk very far to say hello."

"I'm so sorry," said Justin.

"Well, he had a good life. He was eighty-four. He died three years ago so I've had time to get over him."

"And find someone else?"

"Nice, long-term men are hard to find. I suppose I've been spoilt - far more good husbands than any woman has the right to expect."

"How many altogether do you think?"

"Oh, pterosaurus! How long would it take me to work

that out? Some lasted fifty years, some hardly any time at all. Maybe thousands. So, I'm not very used to being alone. Not nice, really. Still, moping does no good, and now I've found some family. I can't wait to see Lekrishta again. A real sister! And three brothers. Wow!"

"Okay then, let's go." He paid the bill and they set off back towards the ferry.

A woman who'd had thousands of husbands. A woman who looked twenty-five, a sweet and innocent cuddly twenty-five. This surely was a nightmare, wasn't it?

"Oh, this gorgeous car!" she purred. "How old is it?"

"No idea," he sighed. "It's not mine. It belongs to Iska, Prince Alexander Romanov. He's promised to murder me if I get a scratch on it."

"Nice man," she laughed. "Didn't he get shot dead by the Bolsheviks?"

"About a hundred years ago. Didn't puncture his ego, did it? Sorry, shouldn't cast nasturtiums on your brother."

"Well, serves him right if he's trying to kill you. Not cricket. That's what you English say, isn't it?"

Somebody in the manager's lodge must have alerted the Great House, for a reception committee waited under the colonnade.

"That's your other brother, Denrico, known to ordinary mortals as Enrico Fermi, the butler."

"The atom bomb man?"

"Apparently. Iska thinks he's trying to create nuclear fusion in his bedroom. Trying for another Nobel Prize."

She burst into peals of laughter.

"All the other people on the estate are ordinary humans and don't know your family are immortal, so we have to watch what we say. Lucrezia has ordered all Immortals to attend every lunch and dinner with her to keep us all

150

in the loop, as she puts it. We're on a war footing, in danger of going bankrupt. That's why we catered for your golden wedding party. We desperately need the money."

"Oh, Higgs Boson! How awful!"

Mrs Lepanto came down the steps to meet them.

"Welcome to Monrosso, Mrs. Wilson. We've prepared the Green Room for you. It has a lovely view over the lake. Please follow me."

Justin handed her suitcase to Denrico and watched him follow her inside. He might lack Iskander's suave flamboyance but he was still, under that sober butler's garb, a tall, dark handsome hunk.

Lucrezia was so devoted to King Cesare that no other man could interest her at all. Kerallyn was so uncannily like her that one of those two near look-alikes seemed sure to capture her. He'd let Lucrezia break his heart. Was Kerallyn going to do the same? Which one would she choose?

With nothing particular to do this Sunday morning, he sat on the stairs in the beautiful entrance hall, admiring the décor, waiting for lunch time and hoping to waylay Kerallyn when she had settled in and came downstairs.

It occurred to Justin that he was still a long way from understanding this strange family. Lucrezia had insisted she tried not to be noticed, so why did they sport such attention-grabbing names? Maybe it was a sort of double bluff. Anyone doubting their authenticity would focus on the names, not their immortality. It seemed a high risk strategy to him. Still, if it was a bad idea you'd think they would have realised that by now. Perhaps being the seventh Marquis of Monrosso was easier than inventing seven different new identities. And these nine hundred hectares on the slopes of the mountains put a protective space between him and the rest of humanity. He could be a self-sufficient hermit if he chose.

151

"What have you done wrong, Sonny?" asked Den, coming back down the stairs. "Who's put you on the naughty step?"

"Born in the wrong village. Waiting to be executed by Iskander, or should I say Prince Alexander Romanov?"

"You still insist you've only been immortal since last Saturday?"

"Yep. Sorry," said Justin. "You lot are twelve thousand years ahead of me. Do you think I ought to dream up a fancy new name like all of you? Why did you pick Enrico Fermi? Why not a duke or something?"

"Never been much interested in titles. You have to do too much fighting to keep them. Science is much more fun. I had more fun inventing gunpowder than the generals get in using it, and nobody gets hurt by my fireworks." He sat down on the steps beside Justin.

"You invented gunpowder?" exclaimed Justin. "Thought that was a Chinaman."

"Yes, my name was Wei Boyang at that time. I wrote out the formula for the Emperor in 142 AD. It was a long time before anyone in the West cottoned on to its uses."

"So, have you always been a scientist?"

"They called us alchemists in the old days, and treated us like magicians. When I was Nicolas Flamel I convinced the silly, credulous creatures I'd created the Philosopher's Stone, that would turn lead into gold, or confer immortality. I did take on titles once or twice. When I called myself The Count of St. German I let people think I was five hundred years old. And I was a prince once, in the seven hundreds, Prince Khalid ibn Yazid. Was offered the throne but let my brother have it so I could get on with my research."

"So why did you plump for a boring name like Fermi?"

"They gave him a Nobel Prize for what he said about the atom. They didn't give me one, did they, back in 500

BC when I first wrote down my insights about the atom? I even invented its name. I'd had all that in my head for thousands of years by then, but nobody could read so there was no point in trying to write it down."

"What was your name, then?"

"Kanada. I lived in Gujarat. If you read what I wrote then it will blow your mind. Everything I described more than two thousand years ago is gospel now: the nature of atoms, molecules, the effects of heat, all there. For some reason, Zosimos of Egypt is credited with writing the first books on alchemy, but no matter - that was me as well."

"So, I still don't understand why you picked the name Fermi. Did you ever try impersonating him in public?"

"Yes, now and then. Did a few lecture tours when the great man was hors de combat. Nice to get a bit of applause, some recognition for my ideas. When he died I packed in as many lectures as I could before it dawned on people that something odd was going on. Maybe I should find another atomic physicist, a very shy one. He could live a double life."

"Why on earth don't you use your own name? You could be famous."

"What name? Denrico's my only real name. We all have to invent identities all the time, at least every eighty or ninety years or so. Might as well pick up an interesting real one if it flutters under your nose, with a ready-made background. We've never had genuine birth certificates - how could we? Even in Mrusha, twelve thousand years ago, we weren't sure who our parents were, and there weren't any registry offices in the Stone Age.

"Nowadays things get worse and worse - computers recording every move you make, everybody double-checking all your personal details. You can't just set up as an alchemist any more. You have to pass exams, flaunt certificates. Seven years slogging away at uni, having to memorise loads of so-called 'facts' I know are nonsense, just to get a doctorate. Then I'd have to work for years in labs, skivvying for dopes who think they know it all. Most of the fashionable theories are just plain bonkers, but they wouldn't listen to reason, would they? It would spoil their precious reputations. They'd just rubbish me. Academia is a shark tank. Just ask Kerallyn. She probably has more patience with idiots than I have. And how Chez and Krish stuck that long course in Medicine I can't imagine."

"Especially if he's as thick as she says."

"Thick? Well, he always manages to get a first class degree, whatever he studies. She's just teasing him."

"Are you really trying to harness nuclear fusion?"

"As far as one can in one's bedroom," he laughed quietly. "Without burning holes in the carpet."

"People say it could cure a lot of the world's problems, including global warming."

"They said the same about atomic fission, but it's proved a tricky tiger to ride, hasn't it?

"Surely Humanity needs you. It seems such a waste."

"Ha ha. They may need me but I'm sure they don't want me. I'm fine here, thank you, doing a lot of thinking and a bit of experimenting in perfect peace. What more could an ancient alchemist wish for?"

CHAPTER 17

A DAY OF REST

Queen Krish had declared that Sunday should be a staff holiday, in recompense for all their hard work the days before.

She seemed to have forgotten that she couldn't cook, and fitted seamlessly in with the rest of the Immortals as they explored the kitchen and piled the big kitchen table up with goodies. There was enough left over from the Golden Wedding to give them all a taste of it for lunch.

"It's pretty good," said Den. "Cook really did us proud."

"They did eat the goat and the alpaca," said Justin. "You were afraid they wouldn't. It's really tasty isn't it?"

"Lucky they didn't get a chance to make a fuss of the alpacas beforehand," said Iska. "I bet that would have put them off their food."

"It's their lovely long eyelashes!" laughed Kerallyn. "So sweet."

"Yes, they do taste quite sweet," grinned Iska.

"Oh, shame!" laughed Kerallyn, "Well, I suppose they have a very good life here, while it lasts. Most food animals get shut up in sheds all their lives these days. Yours have nearly as much freedom as they would in the wild. They all look so alert and confident. They don't even seem very scared of the wolves. That's amazing. It's like a Garden of Eden here."

"Unfortunately we may be in for The Fall in two weeks time, if we can't pay the tax bill," said Justin. "The estate manager absconded when he heard Lucrezia was coming home. He'd somehow run up huge debts in our name. We are likely to be declared bankrupt a week on Friday."

The other three glowered at him and drew breaths as if to remonstrate, then simmered down and looked grimly at their plates. Well, thought Justin, it's easier for me to explain. They must be feeling ashamed of themselves, and as part of the family she needs to know how things stand.

"Oh, stegosaurus! How awful! How much do you need? I can certainly find you a few thousand," said Keralyin. "

"Half a million," said Krish. "Just for the tax bill."

"I've already offered to help," said Justin.

"And we can cough up a bit. Only right," said Den.

"You're already paying the staff's wages," said Krish. "If you wipe yourself out how will you finance somewhere to live if we all end up in the street? And Iska's bought us a few thousand empty bottles for the winery in his own name. We owe the bottle factory so much they refused to supply us any more, so we had tanks full of wine we couldn't bottle."

"It was our fault," said Iska. "Chez and Krish were away in Hell with Medecins Sans Frontieres. We were having a great time hang-gliding while that slimy crook of an estate manager extracted every penny he could out of Monrosso, then scarpered. We were stupid to trust the bastard."

"We all did," said Lucrezia, "He emailed false figures to me every week. Everything looked to be going so well. I didn't suspect a thing."

"He emailed the same figures to us as well, so we were all hoodwinked. But we should have gone down the drive and checked the books and the bank statements for ourselves. But you think, well, he's the financial expert. We're paying him to do all that."

"Well, it's an all too familiar tale, isn't it?" said Kerallyn. "These fraudsters are always so plausible. The way they con very intelligent people out of huge sums of money seems amazing but they do. And you weren't being greedy and asking him to double your money, were you? You just wanted him to do an honest job. I don't think anyone can really blame you, and I think it's a miscarriage of natural justice if you lose this lovely place. You've made it into Paradise. We've got to stop the serpent from ruining things this time. I'm sure you're already doing everything you can, but please tell me how I can help. I'm thrilled to bits that you're treating me as family and I'll pull out all the stops to help. Who's doing what and how can I fit in?"

"Well, Den is trying to turn the Great House into a special events venue. Yesterday was our first attempt at catering. Have we any idea how well it went?"

"From the guests' point of view it was great. Out of this world. They were ringing their friends back in Boston, raving about it and emailing pictures. Your staff looked unbelievable in those fabulous costumes. Mrs Billington tried to call you last night but couldn't get through."

"They've cut our phones off. We owe them thousands."

"How on earth can you run a business without a phone?" asked Kerallyn.

Everyone shrugged in silence.

"We've managed so far using emails and mobiles. We go down to the manager's lodge to hijack the neighbours' Wifi," said Justin.

Kerallyn gasped and shook her head. "You really are up the creek without a paddle, aren't you? I bet the guests assumed you must be billionaires to have a place like this."

"What did Mrs Billington want?" asked Justin.

"Somewhere to hold a party. It's to celebrate their son's graduation. About sixty youngsters. The Castle Hotel was going to do a buffet, but I'd think a barbecue would be better if Wednesday's weather forecast is good."

"This Wednesday?"

"Mmm. Would your staff want to do it?"

"Oh, yes," said Den. "They really enjoyed yesterday. Asked when the next party's going to be. No wasp livery for a barbecue, I think, don't you?"

"Pity, but it would look very odd outside, wouldn't it? Perhaps we could think up something outdoorsy for the staff to wear. They've got a band booked so they could dance if you've got a suitable space outdoors. I've got Mrs Billington's number. Shall I ring her?"

"Come down to the lodge with me this afternoon," said Lucrezia. "Mobile reception is very poor up here without the Wifi hub. There's no phone mast nearby."

"I'll go down to the vineyard," said Iska. "The bottles arrived on Friday, so I'll go get the plant ready to run first thing tomorrow. The brandy's nicely matured. Justin's got orders from a dozen places in Como, so I can go down with the women on Saturday and deliver it while they man the stall."

"Iska is trying to maximise profits from the farm and the vineyard," said Lucrezia. "Justin's full of good ideas and drives around bringing back free stuff like the outdoor tables - "

"And especially valuable stuff like you." smiled Justin.

"We've fixed up solar panels on all the roofs and we've put in a system to turn slurry from the milking parlour into methane gas for heating. We run the Jeep on it too," said Den. "We've cut power costs right out – and the residue is good fertiliser."

"Oh, that's what it is!" exclaimed Lucrezia. "We've got all these strange bills for expensive tanks and fancy valves and boilers. Couldn't think what they were for. We seem to owe a horrific amount for strange gear I don't understand."

"He said he'd got us special EU grants to pay for all

that!" exclaimed Iska. "When the stuff arrived he must have sent the delivery notes off to get the grants, then pocketed the money. And the winery. He got us grants to replace all the old wine vats and bottling machines and replant a whole hillside."

"Somebody's going to get an absolutely first rate farm for damn all when the receivers sell it," moaned Den.

"You could stop spending any more on it and let it go. Then we could raise a load of money and buy it back," said Justin.

"It must be worth 20 million upwards as a stately home estate. How could we raise all that? Loads of billionaire Russians and Arabs and Chinese would fight to snap it up. If we could raise enough to buy it surely we can raise enough to pay the debts instead," said Iska. "Infinitely cheaper."

"Are you offering to try that?" asked Lucrezia.

"Look, Lekrishta, if we were any good at raising money we wouldn't be here now, would we?" Den turned to Kerallyn. "We both got swamped by the market crash in 2008. Lucky we bumped into Chez at a hang-gliding championship and he offered us a home here. Worst thing he ever did, poor soul, wasn't it?"

"Where is Chez?" I haven't seen him yet. Does he look like you two? You looked like triplets in Mrusha days."

"Very similar, yes," said Krish.

"Where is he?"

"Gone away somewhere, last Tuesday," she sighed.

"Gone away? Where? Just left you to sort things out?"

"Twas ever so, wasn't it?" said Krish with another sigh. "The great warrior goes off for months, even years at a time and leaves his lady struggling to keep the peace, to run the castle, the state, the empire, what have you. Then back he comes in triumph and plonks the crown back on his head. The story of my life, and yours, I'll bet."

"Well, yes, though my other halves weren't usually quite so heroic."

"Lucky you!"

"He can't have deserted us. He'll be trying to fix things for us," said Iska."

"How? Finding us a squat?" asked Krish bitterly. "He'd better hurry up. We'll need it in a couple of weeks."

"He's gone hunting again, of course," said Justin. She ought to give the poor man his due. "Rats. Really nasty big rats with nasty great teeth."

"I told him not to -

"Well, what else could he do?" demanded Den. "He's no better at finance than us two. He got out of your hair so you can concentrate, and if he's true to form he'll bring back the bacon, won't he? Then all our worries will be over."

"Where are these nasty rats?" asked Kerallyn.

"Raqqa," said Justin.

There was a sharp intake of breath around the table. Lucrezia put her head in her hands.

"Oh dear!" said Kerallyn.

"Exactly," said Justin. "Oh dear!"

"So, who is sorting out the finances?" asked Kerallyn.

"I'll give you one guess," said Lucrezia bitterly.

"Oh dear!" said Kerallyn.

"Exactly," said Lucrezia. "Oh dear!"

"I'd better make you some coffee then," said Kerallyn, getting up from the table as they all burst into rather hysterical laughter.

CHAPTER 18

MONDAY MONDAY

"Progress report?" asked Lucrezia when the staff had cleared away the pudding plates and Den was pouring the after-lunch coffee.

"Managed to get through to Mrs Billington at last. She's delighted with the idea of a barbecue and she's coming to see Den this afternoon," said Kerallyn.

"We've done 500 bottles of brandy this morning." said Iska. "The lads are keen to help with the barbecue. Gianni wants to dress up as a cowboy, with a red bandana and a ten gallon hat. They could help with clearing up rubbish and fetching and carrying. Wouldn't need any training for that, would they?"

"I've found you some more customers in Tremezzo and Cernobbio. They'd like to try some of everything," said Justin. "You could deliver to Tremezzo on the way to Menaggio and Cernobbio on the way to Como."

"Great. Thanks a lot," said Iska.

"I've emailed the tax demands to Vetrianos in Rome to see if they can make sense of them," said Lucrezia. "I know we have to pay a tax of fifty per cent on the staff paybill, but apparently that's not been paid for nearly two years. It's amazing they didn't send an inspector in. They've used the bogus figures Albertelli emailed us to calculate the income tax, but I think those figures were inflated. Then it looks as if they're trying to charge us inheritance tax. That has to be wrong. They abolished that in 2001. If we'd died before then we'd have had to pay sixty percent on everything we owned. When we realised we could die for nothing we pretty well ran to the death registry. Maybe I should go to Rome, try to get an appointment with a tax inspector, show him our death

161

certificates. Thank goodness they're taxing Monrosso as a farm, not a luxury house, or the property taxes could be millions and we couldn't afford to live here."

"But you've been living abroad for more than ten years so aren't you exempt from taxes anyway?" said Justin.

"That only applies to our personal finances, and the house in Rome. This estate is separate: it's a private farming company. Different rules - and inheritance tax shouldn't affect it anyway. They seem to have got things mixed up, so we could have grounds for an appeal. I hope Vetrianos can advise me about that. And I've had a letter from the bank. I emailed them for print-outs of the estate's accounts for the last three years, but they're saying there's some anomaly they're worried about - surprise, surprise. Asked me to go see them. I'm going this afternoon. Would anyone like a lift to Como."

"I'll take you. Then I can deliver the brandy while you're at the bank," said Iska. "I'll take a few cases of wine as well; see if they'll take that too. Justin's negotiated a good price, far better than the co-operative gives us. If we could sell all of it this way we'd make far more profit."

"You usually deliver to the delis up north on Monday afternoons, don't you?" said Justin. "I could do that and take as much wine as I can get in as well. See if I can get any orders for that."

"Could I come with you and help?" asked Kerallyn.

"You bet!" said Justin.

"I think that's the deli, there, just around the corner. Pull in quick."

The car stopped dead, giving them and the car behind a bit of a shock, and making the bottles rattle. Unlike the old Alfa, this ancient Merc didn't need a heavy hand – or foot. Den had handed him the key without hesitation.

"You're not going to threaten me with sabres at dawn?"

"Mmm? No, she shouldn't give you any trouble. Virtually drives herself. Get cash on delivery, if you can."

Kerallyn came running back and waved him around the corner, then right around the block to the back entrance to the deli. The owner came out to help check the chilled boxes of milk products and prepacked meat and helped carry them inside. Justin went back for a heavy box full of preserves. How long before there would be Manuka honey too, he wondered.

Kerallyn was checking the invoice with the owner. "Yes, two hundred and seventy euros, that's right. You're happy everything you ordered is here?"

The deli owner did a quick check of the preserves. "Yes, all correct. But there's something else I'd like to raise with you - Mrs Albertelli's purchases. She said to send the bills to Monrosso House, but maybe they got lost in the post." He put a pile of bills, all clipped together, onto the counter.

Justin and Kerallyn stared at each other aghast. "How much does she owe you altogether?" asked Justin.

"Two hundred and ninety five euros."

With a deep sigh Justin counted out twenty five euros onto the counter. "Is that it, or do we have any more debts that you know of?"

"That's it. I expect you'd like a receipt."

"Yes. Good to get that sorted out. Don't want to get a bad name, do we? If you get a visit from Mrs Albertelli again I'd advise you to refuse her credit. We will not be responsible for any more of her debts. Do we need to give you that in writing?"

"I've heard tell that the Albertellis don't live at Monrosso any more. Is that right?"

"That's right," said Justin grimly, "so there's no way we'll pay any more of their debts. Now, can we interest you in some of our wines. Absolutely top quality. I've got

163

a few bottles in the car: our special Marquis Reserve brandy, and red, white and prosecco. I see you have a wine section over there. Usually all ours goes for export to top hotels and restaurants, but we could supply you too, as you're right on our doorstep."

Justin hurried out to the car and brought in a bottle of each. The labels looked attractive, a cut above those already on the shelves."Ten euros a bottle for the brandy, as an introductory offer, and five each for the others. Maybe you'd like just ten of each to see how they sell."

"Go on, then, bring them in," smiled the owner.

Nice man, thought Justin. Obviously didn't like to send us home twenty five euros down. So, at least we're two hundred and fifty up, minus my twenty five euros.

"Oh, my goodness!" exclaimed Kerallyn, talking with a woman who had just come in with the owner's wife. "How much does she owe you?"

"Nearly five hundred euros. She bought a coat last winter and a frock and some designer jeans and tops and things. Charged them all up to Monrosso. I sent the bills again and again, but never heard a thing. I hate to ask. I don't want to offend you, My Lady, but I don't make a lot of profit these days and -"

"I'm not the Marchesa. I'm her sister. She'll be horrified to hear you haven't been paid. We must do something about it as soon as we can."

"I don't suppose you'd like some wine on account?" Well, worth a try, he thought. He'd feel a complete twat going back to Monrosso worse than empty-handed. He handed her a bottle. "Very famous high quality stuff."

"Can I ring my brother? He's giving a big party soon and he hasn't decided on the wine yet."

They waited, metaphorically crossing fingers and toes.

"He needs quite a lot. There's to be about a hundred people so he needs about twenty bottles of each, red

164

and white, then twenty prosecco at the start and then again for the toast. That's eighty bottles. Do you have enough?"

"If I haven't I'll go fetch you the rest. I can let you have it at a big discount, as a special deal, since you've been out of pocket for so long. How about five hundred euros and call it quits?" It would cost you a lot more than that in a shop, wouldn't it?"

"So," he said, as they settled back into the old Merc, now considerably lighter than before, "what have we got? Damn all, as far as I can see. We've handed over masses of goodies and come away with a few receipts for Mrs Albertelli's stuff. I could strangle that woman, as well as her husband."

"Well, at least we've saved our good name. But I just hope this isn't going to happen at every shop we go to."

"Lucrezia should be pleased. She says we ought to pay the little people first. The taxman won't suffer if we don't pay but these small businesses could be teetering on the brink."

"Den says we can't afford to be so sentimental, or we'll get busted, then none of the little people will get paid. He wants us to scrape everything we've got together to get the taxman off our backs, then stall as long as we can and try to pay the others when we've earned enough," said Kerallyn.

"Huh! And when will that be? If we'd got paid for the whole car load this afternoon how much would we be taking home? Less than a thousand euros. I doubt if we made much more than twice that on the Golden Wedding. We owe the taxman half a million, then at least the same again for all that eco stuff that Iska ordered, thinking the EC was paying. Everything we earn will be a drop in the ocean. Chez is our only hope."

"What exactly is he doing? Nobody will talk about it."

"Nobody likes to think about it."

"Giant rats. Is he down the sewers, all thick stinking grease and excrement? Oh, yuk!"

"No, it's not sewer rats he's after. He was very quick a couple of weeks ago. Found the chief rat in only a day or two. He said he told them all he was an arms dealer, so they came looking for him trying to buy stuff. This time he's hoping to get work in the new medical school in Raqqa. I can't see how that's going to work, can you? If he does get the job he'll surely have to work all day so he won't have much time for rat hunting. And how will he find them? They won't come looking for him, will they? It could take him weeks, months."

"Why didn't he go as an arms dealer again, if it worked so well?"

"He felt bad about doing it this time and keeping the money for himself, so perhaps being a doctor not an arms dealer made him feel better about himself."

"Doing what? He surely can't feel bad about killing nasty big rats."

"Terrorist rats. Our side bomb the fanatics' camps and hideouts but the real kingpins are too clever to get caught that way. They have their headquarters deep inside places like children's hospitals. Sometimes even squads of Navy Seals can't get at them, so they put prices on their heads. Chez finds an excuse to walk in unarmed and knocks the rat's head off. Then the rats' bodyguard give him a hard time, of course, so he tries to wind them up so they kill him quickly. Poor Lucrezia tortures herself imagining the horrible things that must be happening to him."

Kerallyn groaned and shuddered. "It's too horrible to think about. No wonder Krish is so upset. Poor Chez. How many rats will he have to kill?"

"They've given him three names. They pay millions of dollars each for the real kingpins. Chez felt he could justify it when he gave all the money to the field hospitals and the refugee camps. And, of course, he believes the world will be a better place without those evil murderers."

"So, while we make hay in this lovely place, he may be going through hell," she sighed. "Poor Chez!"

CHAPTER 19

WHAT BANK ACCOUNT?

"Oh woe, woe and thrice woe!" moaned Den, after they had all recounted the disasters of their afternoons.

They all tried to laugh, but it didn't ring true.

Lucrezia waited until they had finished dinner before ruining their appetite with the worst news of all: the errant manager appeared to have closed their Popolare bank account.

"Closed it? When? Do they still have a record of our transactions? Which bank did he transfer us to?" They all gabbled at once.

"All I can get out of them is that we don't have a bank account so they don't have any records of any of our transactions at all."

"You mean the bank just let him close it without our permission."

"They won't even admit we ever had an account with them."

There was a stunned and puzzled silence for a while.

"Are you sure you went to the right bank?"

Lucrezia gave him a withering look, then shrugged and sighed. "Am I going crazy? Surely we've always banked with the Popolare, haven't we? The one near the Piazza del Popolo. I can remember the name of the manager. He used to live near here, so we had him to dinner now and then. His name seems to mean nothing to them. It feels like a nightmare. I can't get my head around it."

"When did any of you last go to that bank?" asked Justin.

"My bank's got a branch in Cernobbio," said Den.

"And mine too. Never been to a Como bank," said Iska.

"It must be at least five years ago," said Lucrezia.

"Bank staff change so often these days. They could have had a couple of new managers in that time," said Justin. "If you're sure it's the right bank we have to think this thing through. Think how that crook managed to do it, then find a way to pressure the bank into investigating what happened, where your money went to. They'll try not to admit responsibility in case you try to sue them."

"You could try to find your friend the manager. Maybe he'd help," said Kerallyn. "Can you remember where he lived?"

"Yes, good idea. I'll try to find him on the net, get his phone number. If he was still manager at the time I'm amazed he let it happen."

"Maybe he was away at the time or something. Albertelli couldn't just close it, surely, and stuff a million into his pockets. He must have transferred it to some other bank. Surely he couldn't do that without your signature. Who had the right to sign your cheques? Who did the statements go to?" asked Justin.

"Albertelli," muttered Den and Iska.

"Didn't you have to countersign them?" asked Justin.

"We did at first, then when we were going to New Zealand for that big championship he threw a fit, said how could he operate while we were away and what if we broke our necks and never came back?"

"Have the police any idea where he's gone to?" asked Justin.

"We don't want the Police involved!" exclaimed all the Immortals in unison.

"Haven't you reported him yet? Why ever not?"

"Engage brain, you dope!" said Iska. "Tell me, Sir, aren't you the man who died hang-gliding in the Dolomites last year? What's your name and date of birth? Who are your parents? Where are you from? Who was your last employer? Were you dismissed for financial misconduct? Well, it would be easy to construct a case against us, wouldn't it? We put Albertelli up to it, then we follow him out and share the loot, half a million each."

"So, did you?" asked Kerallyn, looking him fiercely in the eye, then turning to stare at Den.

They almost fell off their chairs with spluttering outrage.

"Look here, you," growled Den, "If we were so crooked we wouldn't be daft enough to stay on here spending our own money trying to keep the place afloat, would we?"

"And where are we going to find another home a tenth as good as this?" asked Iska. "You don't pick up homes like this for half a million, do you? If I were turning criminal the payout would need to be fifty times as much as that."

"So, we know your price, then," smirked Lucrezia. She caught the napkin Iska threw at her and handed it back to him. "We need to think if he could harm us through this new account he must have opened."

"Maybe the police could pressure him or your bank into revealing which bank he's transferred it to," said Justin.

"You really must be new to this game if you don't know how essential it is to stay under everybody's radar. Getting the police involved could be catastrophic. We have to deal with this ourselves," said Den. "No point in trying to find him. The money's probably gone to the betting shop by now. If he thought we were going to beat him up he'd run snivelling to the police and invent sordid things to accuse us of. Mud always sticks, especially if you have a high profile."

"He might try raising funds in your name," said Justin. "Maybe you should advertise that you disown anything he tries to do. You've got lawyers in Rome, haven't you?

Why not see what they suggest," said Justin.

"Yes, maybe I'd better do that tomorrow morning."

"I worked in a bank once," said Kerallyn."Confidentiality was very important. We were banned from revealing anything about our clients. Maybe he hasn't actually closed the account. There'd be no need if it was in his sole name already, would there? Am I right in thinking that your name wasn't on the account, Krish? If so, then you don't have an account at that bank, do you? They can't give you info about somebody else's account."

Everybody groaned and sighed.

"Idiots, we are!" wailed Iska. "We just handed it to him on a plate. How could we be so utterly brain dead? Now what do we do?"

"Well," said Justin, "if you did have an internet account at that bank, I might be able to sneak in and have a look around. Depends how good their online security is."

"Of course!" exclaimed Kerallyn. "You're a computer expert, aren't you? You can hack your way in."

"I can try. Can't promise, but I'll give it a go. Somebody will have to go open an online account tomorrow to get me details of their system, then maybe we can find out what he's up to."

There was a huge sigh of relief around the table.

"So, said Lucrezia, "tomorrow I go open a new account with the Golden Wedding cheque. And we'd better think if any of our money might still be going into Albertelli's account automatically and get it stopped. The money from the Co-operative must be, and the tenant farmers. Must get in touch with them as soon as I've opened the new account. And his wages must be, I suppose."

"Can't be, unless he's paying himself out of the account he's stolen. We don't have an account to pay him out of. The staff said he'd already stopped paying their wages," said Den. "Must have been on the point of doing a bunk even before he knew you were coming. Now, I can't see what more we can do about the bank tonight, so can we think about the barbecue? It's the day after tomorrow."

"Oh, supercollider quarks!" exclaimed Kerallyn. "Can we do it in so little time?"

"Got to," said Den. "Mrs Billington came this afternoon and we agreed the whole thing. Fifty euros a head, including all the drinks. They're hiring coaches so they can have a few drinks. That should get rid of the parking problem. Now, Prince Alexander, what are you going to do about your blasted animals? We've agreed to use the grass in front of the House for the party, but how are you going to stop your rampaging monsters from trampling everybody to death?"

"It's just the novelty. They'll get bored eventually, and anyway, I'm training the wolves to work them. Can't afford passengers now on this estate, can we? They'll have to work for their keep like the rest of us."

"Can you really do that – train a wolf to act like a sheep dog? Surely it's their instinct to kill the animals."

"We've been doing that for tens of thousands of years. They don't act like dogs, they *are* dogs. All the so-called dogs are just weird mutations, but they're all ninety-nine percent wolf if you see their DNA. As for the instinct to kill, look who's talking. Man's the worst killer of all."

"Well, good luck. It should be exciting to watch," said Kerallyn.

"I think you need a haha," said Justin."Much better than flimsy stakes and wire."

"We've got a haha," said Lucrezia. "Capability Brown insisted we must have one. We couldn't see the point. We wanted the animals to graze right up to the house. Why have to scythe it when they're keen to eat it? But I can see the point of it now. Just what we need."

"Where is it?" asked Den. "I've never noticed it."

172

"Just where it should be, about fifty meters from the front of the house."

"You mean that little wall? It's only a few bricks tall."

"Well, it was taller than me when it was built, so how has it managed to shrink?"

"The men dump all the spare soil and compost there. They say it's what everybody's been doing for yonks, so the animals can get up to eat the grass on top,"said Iska.

"You could hire a digger and dig it out then," said Justin. "It must be quite soft."

"Got a nice new digger. EU grant," said Iska, and everybody huffed sardonically.

"Could you get that done?" asked Lucrezia.

"As you wish, My Lady," said Iska. "We'll get onto it first thing tomorrow. Have to think where to put the spoil. Now, what about the cowboy costumes for the lads?"

"I'll see what I can find you," said Justin.

"What about the food?" asked Lucrezia.

"Maria and Cook and I have been sorting that out all afternoon," said Den. "The wives are coming to help in the kitchen and we've managed to get the chef from the Castle Hotel to help Cook do the barbecue."

"So, full speed ahead,"said Lucrezia."Let's hope it goes as well as the first one. Should be very different."

"I've spotted a sale in Como tomorrow," said Justin. "Bankrupt stock. If you can think of anything you need I could go along."

"We've got a large estate stuffed with bankrupt stock already,"growled Den."Why would we want some more?"

"What about your farm shop? Where are you going to site it? Have you got display cabinets and stuff?"

"We're going to put it in the dairy, out on tables, we thought," said Iska.

"Won't the customers be a danger to the livestock?" asked Den. "And how will you know how many veggies to pick? They'll surely wilt if you just put them on a table. Won't the food inspectors take a dim view of that? They'll insist you have proper chilled display cabinets."

"I think Den may be right," said Lucrezia. "If we site the shop up here in the garden they can point to what they'd like and we could cut it fresh for them. They can use the car park up here then, so they won't be mowing down the animals in the dairy yard, will they?"

They were still arguing the pros and cons as they took their drinks out onto the colonnade.

"Maybe we should leave the residents to battle it out, don't you think?" Justin whispered to Kerellyn. "Would you care for a short walk in the moonlight?"

"Mm, lovely idea," she murmured. "Oh, but what about the wolves?"

"Maybe you're right," he sighed. Those blasted wolves again!

CHAPTER 20

SMELLY OLD SHOPPING

"He's as good as his word, isn't he?" said Justin. They could see the new digger heading for the submerged haha as they drove down towards the gates "Your family are pretty useful when they set their minds to it."

"Well, we've had quite a long time to hone a few skills," she smiled. "Trouble is, though, we've done everything so many times it gets to be a bore. Sometimes you feel like curling up in a hole and going into a thousand year sulk, just to get rid of a bit of time."

"I overheard Chez and Lucrezia saying exactly that a couple of weeks ago. They'd both had a bad time in the Middle East."

"How long have you known them?"

"Oh, about three weeks – Lucrezia, that is. Chez only about five days, so hardly any time at all. I suppose you've known them for twelve thousand years."

"Oh, no! After Mrusha was destroyed we lost contact within a few months, right until the Golden Wedding party, I can't claim to know them at all. We must all have changed completely in twelve thousand years. I can't understand how I recognised Lekrishta. We were both dark skinned – we all were in Mrushan days. I suppose we must all have lived a long time way north of the Equator."

"Look at that! A bit premature, surely."

Two farm hands were positioning a big notice board onto the railings outside the gate under Iska's directions.

"Well, we'll have to buck up and get the fridges somehow. Let's hope there's something useful at the sale." The bottles rattled as he swung the old Merc out onto the Como road.

"It really is so beautiful," sighed Kerallyn, as they drove along the lake shore. "Hugh would have loved it."

"Hugh?"

"My late husband, Hubert. He loved to see the sun sparkling on the water. He went blind in the end." Her voice choked a little.

"I'm so sorry."

"I though I'd put him well behind me. Can't cry for thousands of husbands, can I? What a horrible cry-baby I would be. Let's think of something cheerful instead."

"Tell me about Mrusha. Everybody keeps insisting I must remember it, but I can't believe I've ever been there. What exactly was it?"

"Just a village. Maybe it was freakish for its time. I never came across another village for maybe a thousand years. There were twenty or so big huts, big enough for at least ten people, and loads of little ones where a couple could spend a bit of time alone together."

"So, maybe three hundred people?"

"Mm. More like four or five hundred. Hard to remember. It was in a perfect spot, with forest behind and a huge lake in front. We had unlimited clean water and, of course, loads of fish. The young men went into the forest and came back with meat, and the children picked fruit and veggies. The women got the idea of keeping animals we'd caught alive until we needed them, and they dug up plants and replanted them where they grew fatter and riper right under our noses.

"That really peeved the other tribes who didn't have a settled home. They kept raiding us, trying to steal our animals. They seemed to think that keeping animals and

176

plants as captives was a crime, offended their gods. Our young men had to keep defending us, so we chose Chez as our war leader. He was great, would take on any enemy. Everybody adored him. He never let us down.

"Then, when it had been lovely and peaceful for ages, hundreds of Skunduns suddenly appeared from nowhere in the middle of the night. They rampaged right through the village, hacked everybody to pieces, then they burned Mrusha out of existence.

"Next morning, we got the shock of our lives. Something we can never forget. A few of us woke up in perfect health. We couldn't understand it. Felt frightened out of our wits. We could all remember being butchered by the Skunduns, but all our wounds had disappeared overnight. We went looking for other survivors, but everybody else was dead or dying. Just thirty of us survived, but in perfect condition: not a single cut or blemish, unreal, like waxworks. It was so odd it still gives me the shivvers to think about it."

"Even though it's now happened to you thousands of times?"

"Yep, even now it gives me goose bumps. How could you possibly have forgotten Mrusha?"

"That's what Iska says. I can't possibly be an Immortal, can I?"

"Well, I can't see any reason why not. Why should we thirty be the only ones? Whatever quirk of evolution created us could surely happen in some other place besides Mrusha, or in some other era. You could be part of a new modern strain, newly mutated, couldn't you?"

"Well, it's never happened to me before," said Justin.

"There has to be a first time, doesn't there? The Skundun Raid was the first time for us."

"Phew!"breathed Justin, goose bumps rising.Could she be right? Was this a blessing or a curse? It certainly felt threatening, like standing on cracked and creaking ice.

"I'm so lucky to have met you," she said.

"Really? Well, I feel very lucky to have met you too." Another gorgeous Lucrezia, this time mourning only a dead old human husband, not an absent immortal lover. Was he in with a chance this time?

There was a layby ahead, a place where tourists often stopped to film the view. With a clink of bottles he slewed quickly into it. "Here you are," he said. "What would Hubert think of this?"

She gazed out over the gleaming water, over the shimmering reflections of the bushy green mountains, to the blue peaks way beyond. She drew a deep breath and sighed. "He'd have loved it. For the last two years I'd have had to describe it to him." Her voice broke and tears began to trickle down her cheeks.

He pulled a tissue from the door bin, pushed it into her hand and pulled her head onto his shoulder. At last he could stroke her lovely glowing hair. It felt exactly the same as Lucrezia's. She'd been crying too.

"I'm so sorry to embarrass you like this" she snuffled.

"No problem, it's my speciality," he murmured. "Providing a shoulder for gorgeous women to cry on - about other men."

"You're so sweet. You're gorgeous. I can't imagine you ever making a woman cry."

"No, it's usually the other way round," he murmured.

"Oh, how sad! Now I want to cry for you."

"I can think of something more effective you could try."

"Mmm?" she asked.

"You could kiss me better – unless you think I'm too repulsive."

He leaned towards her slowly and she didn't back away. She let him kiss her gently, then gently kissed him back.

"How could anyone think you repulsive," she murmured, and kissed him gently again.

Oh! Wow! thought Justin. Words fail me. She is absolutely !! He slid his arms around her and hugged her like a comfort blanket.

"You're so comforting," she murmured.

"I expect I fall far short of Hubert, but I'll do my best."

"You have one huge advantage over poor Hubert," she said. "He's permanently dead. You're permanently alive."

There was a squeal of brakes and a scrunch of stones and a police car swung into the layby in front of them.

"To be continued?" murmured Justin, as he turned the ignition key.

"At your service, Sir," she smiled.

A large warehouse full of bankrupt stock is not a pretty sight, especially if your own bankruptcy is looming. It was obviously the best place to shop in the circumstances, and they soon felt surprisingly lucky. Three refrigerated display cabinets, stained and smelly but intact and labelled "working", were knocked down to them at a very good price.

"Here's hoping they're not back here in a few weeks' time," said Justin."Now, fancy dress and bits and pieces."

"That red stuff. Let's try for that," urged Kerallyn. "I looked in the box. It's mainly picnic stuff. Should look nice and cheerful."

It was a long wait until the fancy dress came under the hammer, and by then most of the customers had left. The auctioneer was clearly tired and ready for his lunch. Impatiently he knocked down bundles of grubby

179

crumpled clothing as soon as anyone showed enough interest to make a bid, however small. Justin threw caution to the winds and bought nearly all of it. Who knows what might fit the bill if Monrosso survived long enough to need it. Probably money down the drain, he thought, as he offered his new debit card, but he could well afford it.

They drove back in triumph with the old Merc stuffed to the gills. The bottles in the boot no longer clinked, smothered by booty. The back seat was full right up to the roof, and topped off with big hats of all descriptions. Thank goodness for the wing mirrors! Despite the air conditioning, the musty smell was pretty noxious. Somebody would have to do a lot of washing.

CHAPTER 21

FUN IN THE DARK

Wednesday had much in common with a French farce: everybody chasing around getting in everybody else's way and wailing that nothing was going right.

The banner across the new farm shop notice board read: OPENING SHORTLY. It might as well have read: COME IN NOW, as that's what three car-loads of people did. They followed Iska's signs to the dairy, just as Gianni was hosing the cow-clap towards the drain, with half a dozen cows harassing him, trying to get an unscheduled shower. The visitors got their cars splashed and their feet sticky. One even got a friendly shove from one of the beasts. Not a good start for the shop.

Gianni dropped the hose and took them into the dairy. The wives, in their hygienic white overalls, caps and gloves, were busy packaging cheeses and bottling jams. They tried to keep the visitors near the entrance, where they wouldn't contaminate the produce, but of course it was hopeless. There were only two wives versus eight visitors, who were soon sticking their fingers into everything. The food inspectors would have thrown a fit.

"Don't you have any fruit and vegetables?" demanded one of the visitors. "Not much of a farm shop."

"We're not really open for business yet. The garden's up by the Great House, about half a kilometre away."

Iska arrived just in time to take control, sell the visitors some steaks and cheeses, and shepherd them up to the kitchen garden.

Justin and Kerallyn, discussing the relative merits of the orangery and the potting shed, watched as Iska picked fruit and vegetables for the customers and

181

tumbled everything together into a few grubby used plastic bags.

"The shop is going to be up here, in the garden. We look forward to seeing you again, when we're open," he said, shepherding them towards their cars.

"Good," said Kerallyn. "He's learned the hard way."

"The orangery's no good - too hot," said Iska. "Have to clean out the potting shed and use that till we've time to build a proper shop. Don't need to do any potting for the next few weeks."

"You could make the orangery into a nice little café," said Kerallyn. "Lovely and warm in the winter."

"The cabinets are coming this morning, so we'd better get the potting shed ready," said Justin. "Where do you want all the pots and stuff?"

"Have to go in that corner of the garden for now."

"Progress report?" said Queen Krish, as they gobbled down their lunch.

"The shop's going to be in the potting shed for now. Chiller cabinets have arrived. Just need an electrician to get them going and someone to clean them up," said Justin. "We've already had customers this morning, so - "

"The party doesn't start till seven-thirty, so I've time to fix them this afternoon. The girls can clean them out," said Den. "The wives are coming up at six o'clock to help with the barbecue. Then we'll all have to lend a hand."

"The men are doing the milking tonight so the lads can come up and help. Any luck with the cowboy hats, Justin," asked Iska.

"Yip, reckon so,"smirked Justin."And chaps with fringes and leather waistcoats, one with a sheriff's star on."

"I've got us a show as well," said Lucrezia. "I called in to tell the riding school to move their standing order to me. They were still paying their rent to Albertelli, of course. They offered to do a little torch display for us about eleven, when it's good and dark. If we could catch our horses we could join in."

By eleven the party was little short of a riot. Sixty-odd graduating students in celebratory mood were no mean feat to handle. Luckily it was a very warm night, so lots of them, stuffed with food, had collapsed, almost comatose, on the grass, enjoying the music, but the roaring drunks were – well, roaring. Arm in arm they staggered about, shouting and looking for trouble. Bit like the goats, but less charming, thought Justin. Pity I suggested the haha. Might have been fun to see these louts being shoved and butted by the livestock.

"You were right to suggest putting the House out of bounds," he said to Kerallyn. "Thank goodness the Portaloos came in time."

"I've just spent five years at Harvard, so I know what youngsters can get up to. They've paid for a barbecue and a kegger, not a chance to trash a stately home."

"Kegger?"

"Lots of brewskis, a beer-up. Oh, look!"

As if they'd read her mind, a gang was heading up the steps. Suddenly they changed their minds so abruptly that one fell down the steps, bowling another one over.

"F***ing hell, wolves! It's true. They really have got wolves!"

Five wolves detached their evil jaws from the bones they were crunching and emerged from behind the columns, snarling and wagging their tails. A howl came from someone in the crowd and the wolves set up a thrilling chorus.

"The cabaret," laughed Kerallyn. "Still only five. Has the rogue one gone to the vet's?"

183

"Nobody will say. I guess Iska's done it in, instead of me."

Suddenly the band struck up the music from the 'Horse of the Year Show.' From behind the House the cavalry came trotting, carrying burning torches. They processed to the grass below the haha and began weaving flaming patterns in the dark.

A few of the guests managed to walk, but most were so sozzled they staggered or crawled to the edge of the grass and dangled their legs over the haha.

It was a magical sight: horses barely visible in the inky darkness; riders' faces eerily lit by flickering flames. Beyond the great ghostly cedars the lake glittered darkly through the trees.

The magic was lost on a few, of course.

"What about you, all dressed up like a rhinestone cowboy? Bet you've never been near a cow, or a horse," yelled a reeling drunk. Gianni pretended not to hear and patiently picked up rubbish from around his feet.

"Yea, let's see you up on a horse," yelled a big rugby jock, sticking his great beery face right up close to Iska's.

"Sabres at dawn," muttered Justin to Kerallyn "That's what he threatened me with if I messed up his car."

Iska beckoned to Gianni, then pointed into the gloom. The animals, predictably, were watching from the shadows. He whistled. A dapple grey stallion emerged from the darkness and strode towards him. "Call Rupo, Gianni. I think he's there somewhere."

Gianni whistled and called, and drew a whole posse of horses from the blackness, all au nature, not a bit or a saddle amongst them. As the grey drew level with the haha, Iska stepped down onto its back and steered it away from the wall with his legs. The grey danced about, shifting Iska's weight onto a more comfortable position and turning its head to greet him.

"There's Flossie, look!" said Den. "Showing off, trying to get your attention. Go on, give her a treat."

"Flossie, Flossie! Come here, Sweetheart," called Lucrezia. "Fludo, come give Den a ride. Come on, Den, he's longing to join in. Justin and Kerall will keep an eye on things here, wont you?"

"No peace for the wicked," groaned Den. "He'll probably throw me, the tricky old beast. He hates having anyone but Chez on his back. Don't you dare tip me off." He wagged his finger at the powerful horse. "Or I swear I'll make burgers out of you. Yes, I could, so don't you forget it." He stepped down onto its bare back as it stood as still as a statue. "I don't believe it. You're just trying to catch me off guard."

The huge black stallion suddenly bucked and kicked and wriggled, then set off at a gallop towards the torch bearers.

Looks like an advert for Lloyds Bank, thought Justin, Poor old Den, in for a broken neck, for sure.

Fludo came thundering back, with Den slapping his rump and yelling, "Faster, you lazy beast, faster."

"He has hidden depths, this new brother of mine," said Kerallyn. "Puts on this butter wouldn't melt in my mouth act but he's really as tough as they come. Well, why am I surprised? He has Cesare's DNA."

Iska spotted Lucrezia. He galloped after her, leaned over and dragged her off her horse. He ducked his head and straightened up with her sitting on his shoulders.

Den galloped up alongside, reached out and grabbed her. As their horses thundered together past the haha she stood with a foot on each of their shoulders.

"Crikey Moses! Can you do tricks like that?" Justin demanded.

"Never thought to try," said Kerallyn. "She's crazy, isn't she?"

"You can't ride a horse, then?"

"Of course I can. It was the most convenient form of transport till a hundred years or so ago, but unlike some people I never saw it as a sport. She must have practised lots to be able to do stunts like that. Far more adventurous than me."

"I've only known her three weeks," said Justin, "and in that time I've seen her leaping across the rooftops in Rome and jumping off the Burj Khalifa in Dubai. She must be trying to keep up with Cesare. Did the two of them once work in a circus, do you suppose?"

"They must have done, mustn't they, running over rooftops and jumping off high towers. I expect they did a trapeze act or walked the high wire."

"She walked the prop that stops my house leaning onto the house next door. It's five floors up and only ten centimetres wide."

"Better her than me!" she laughed.

"And Cesare must have done a knife-throwing act. The first time I saw him he whizzed a knife at me – missed me by a whisker. I wouldn't be surprised if she played the lady in his knife act. I think she's rash enough. She deliberately provoked the Rome police to chase her in her McLaren P1. She gave them the slip by driving through the ancient Roman sewers."

"If you like women as daring as that you'd find me sadly lacking, I'm afraid."

"Well, that's a relief," he smiled. "If you like men as daring as your brothers you'd find me sadly lacking. Are they really your brothers? You don't look at all like them. Cesare implied that Lucrezia's not really his sister."

"Well, we had no concept of family in Mrusha. Everybody slept with whoever they felt attracted to at the moment, so nobody could be certain who their father was - or their mother. All the women fed any baby that

seemed hungry. We didn't have any concept of owner-
ship. If you needed it you just helped yourself to it. No
one had the right to complain. That was the theory, but
human nature being what it is, well - "

"There were quarrels?"

"Yes. That's why we needed a queen: to organise
things and decide what was fair. With Krish as Queen
and Chez as our war champion everything seemed rosy;
until the Skunduns ruined everything. There were
hundreds of them, firing blazing arrows, set our straw
roofs alight. Butchered every one of us."

"And when you woke up immortal, what did you do?"

"Well, for a while the thirty of us stayed together, tried
to find a safe place to build a new village, but there were
too many Skunduns around. They seemed determined to
wipe us out. When they chased us we kept being split up
and lost. Then the Skunduns captured me, and that was
that. Lost touch with my people until last Saturday."

"What happened when they captured you? That must
have been terrifying."

"Well, I was pretty scared for the first few days, but
they treated me like a goddess, gave me the best they
had, so actually it wasn't bad at all. I just had to make
the best of it."

Lucrezia came galloping back and slipped off her
horse. "Thank you, Flossie, you gorgeous creature.
Good Night." She gave it a hug, then slapped its rump
very gently, and watched it disappear into the darkness.

"Time to go home, do you think? It's past midnight. It's
a barbecue, not a nightclub. What do you think?"

"The bus drivers said midnight," said Justin. "Why don't
we ask the band to play 'Auld Lang Syne' and then just
shove them out? They're so drunk they won't remember
anything tomorrow morning,"

In the end the five men had to link arms and try to

sweep the unwilling stragglers towards the buses.

"Who do you think you are, pushing me around? I'll go when it suits me. Just you shut your face and get me another beer," leered the rugger hunk.

"Yer, you heard. Get me a beer!" hiccuped another.

Iska whistled for the wolves. They came bouncing down the steps and rounded up the drunks like a flock of alpacas, diving at their ankles and giving them a nip.

"Auooo!" called Iska, pointing towards the buses. The wolves joined in the chorus, and drove the cursing stragglers to the buses.

The house maids, disguised as cowgirls, seemed to have found themselves instant boyfriends, and wailed as the men tore themselves away.

" What a night!" breathed Lucrezia, as they watched the tail lights heading for the gates. "Thank you, everybody, for all your hard work."

"Thank you for letting us join in, My Lady," said one of the farm hands, with an arm around his wife. "It's ages since we had a chance to dance together."

"Well, it was a bit of fun, wasn't it?" said Iska, looking at the lads.

"It was great, Mr Romanov. When are we going to do it again?" asked Luigi.

From the door of the Green Room they could see the magical view.

"Gorgeous!" said Kerallyn. "I must have the best room in the House."

"No, I think mine is even better. Come and look." He led her to the room next door. "Voilà! How about this?"

"Don't put the light on," said Kerallyn. "It's so beautiful I can't bear to spoil it." She walked over to the window.

"Even better then with the door shut?" Justin ventured.

He closed the door to shut out the corridor light and joined her at the window. The moon had risen now and the cedars were sharp black silhouettes against the gleaming lake.

"Paradise!" he breathed.

"Only three more days in The Garden of Eden, then I'll be cast out into the great big horrible world again."

"The court case is not till the end of next week," he said. "And even then - "

"I'm due to start work on Monday. They expect me to leave here before then."

"What date is next Monday?"

"September the first."

Justin groaned. "You're right. And I've a meeting on Monday. Pretty important one. Have to get back to Rome by Sunday night. Horrible thought."

"Poor waifs and strays we'll be," she wailed.

"We'll have to comfort each other," he murmured, "then maybe it won't be so bad." He slid an arm tentatively around her shoulders and gave her a gentle squeeze. She reached up and gently caressed his hand. His heart gave an excited flip. Was this an invitation? She'd let him comfort her when she was crying for her husband, but so had Lucrezia. He'd misread Lucrezia and paid dearly for it. Was he falling into the same trap all over again?

Have to take the plunge. Could there ever be a better time than this?

She didn't try to pull away - far from it: as he turned towards her she turned too. The kiss was mutual. Had Lucrezia ever kissed him quite like this? He had certainly kissed Lucrezia, several times, but had she ever returned a kiss? Probably not. At best she had tolerated it - just enough to beat him at her cheating game.

"Do you play games - like Sabrina Fair?" he ventured, between kisses.

"Never heard of it. I'm not terribly good at games. I quite like Snakes and Ladders. Would you rather we played that?"

"No I jolly well wouldn't. Hugging you feels the best game in the world." He kissed her once again. Hell, if she was going to break his heart let her get on with it. Enjoy it while it lasts. "But what about Strip Poker?" he asked, gently pushing off her cowgirl waistcoat.

"Could you teach me? I don't know the rules - "

"Your turn now. See if you can get my waistcoat off."

"There you are," she said. "Now what?"

"I have to try to get your skirt off – and now - "

"I can't take your skirt off. You're not wearing one."

"You have to improvise. Easier for you if I sit on the bed." He had to help her as she struggled to drag off his pants. They were both shaking with laughter. From then on it was a free for all, a contest to see how quickly they could strip each other bare.

Monrosso beds were soft and cosy and Kerallyn was a perfect match, soft, cosy and delicious beyond belief. And those thousands of husbands seemed to have taught her things he'd never dreamed of. If she was playing Sabrina Fair the score was definitely one nil to him. Crikey Moses! Holey Moley! If this is just a night-mare let me never, never wake up again.

He did wake up, alerted by a knock on the door. Kerallyn must have heard the door handle turn because she pulled the sheet over her head.

"Hello?" called Justin.

"Your morning tea, Sir, Madam, My Lady."

"Den! What on earth are you up to?" he growled, as

190

Den, in full butler gear, brought the tea tray to the bedside table, beautifully laid, and with two fine china cups.

"Mr Fermi, Sir, if you don't mind. You surely do not mean to reprimand me for simply doing my job."

"It's the first time you've done this in the whole time I've been here."

"I could say the same to you, Sir, but I would not be so presumptuous." He lifted the sheet a little and pulled out a strand of her hair. "Good morning, Madam," he said. "It seems I owe Prince Alexander a big tureen of borscht."

CHAPTER 22

HACKING

"The potting shed looks disgusting. No hygiene inspector's going to pass that," growled Den over breakfast. "We'll have to build a proper shop."

"Well, obviously, if we stay in business long enough. But it's stupid to turn customers away. We need the money." said Iska. "Local people are probably curious, dying to come in and look around."

"Why don't we whitewash it?" suggested Justin. "That should transform it completely. We've nothing else on today, have we?"

"Speak for yourself," growled Den."I'm trying to find the time to build a mobile phone mast on the Great House roof. The Four G receiver's arrived, but I can't find the time to build the tower and wire it up. It's a blight having to go down to the Lodge all the time and reception down there is very patchy."

"We've a women's club tea party tomorrow and a silver wedding on Saturday evening,"said Lucrezia."but nothing today. Can any of you spare the time? I need to contact the co-operative. If they usually pay us at the end of the month the money will go straight to Albertelli, so I'd better go down there and phone them quickly. You were going to try hacking into the bank, weren't you, Justin?"

"So, while you two go to the Lodge the rest of us had better do the potting shed," said Den. "We should pull the cabinets out again and hose the place down first, blast the spiders out of the roof. In this heat it may dry by this afternoon."

"I seem to be stymied at every turn," Lucrezia sighed,

when they reached the Lodge. "The Banco Popolare refused to open me an account yesterday because I don't have any identification. My passport and driving licence all got burned. I had to fall back on my Rome bank. They've opened me a new account just for Monrosso. They'd just sent me a new debit card, so they accepted that as proof of identity. At least I've got something to pay our earnings into now, but it won't help you with the hacking, will it?"

"Hmm, let's think," said Justin. "Those standing orders from your tenants were going into Albertelli's account. Maybe they could lead you to the account number."

Half an hour later the riding school rang back. They had found a cheque from Albertelli. They hadn't been able to pay it into their bank as it was incorrectly filled in, but, of course, it had his account number and his name printed on it. A big step in the right direction. Now all they needed was the password and pin number.

It was a good ten years since Justin had attempted anything so nefarious, and by lunch time he had only cracked the pin number – easy with only four digits – but the password was a different matter. Maybe he was too old, no longer a sharp-witted chippy teenager, out to cock a snook at the world. Long ago, while still sulking about his expulsion from school, he had designed algorithms to crack passwords and pin numbers, but they were all hidden away in Rome, in secret drawers.

Unlike him, the potting shed team were pretty smug by lunchtime. They had washed down the potting shed and the orangery, untying the wall creepers and laying them on the floor. They'd found a big tin of whitewash and a spray gun. It seemed they'd been brain-storming as well.

"We could run adventure courses in the woods, have paint ball wars and zip wires," said Iska. "We could link up with the riding school and do some really hairy stuff, paint ball wars at full gallop."

"We could get rid of your animals and have a golf course," said Den, grinning at Iska's filthy look. "And we could fit a wire tow and launch gliders."

"Do we have a long enough flat stretch for landing, or could you land uphill?" asked Justin.

"Spose so. With the right plane. When the wind was too tricky we could always land at Lugano. It's only a few minutes away," said Den.

"Rabbits!" enthused Kerallyn.

"Rabbit stew?" asked Lucrezia, "We could ask Cook - "

"No, angora rabbits. Saw them on TV. One rabbit's enough for four sweaters. Honestly, that's what they said, and they're so cute. And chinchillas. We could have a petting zoo. All we need is a few little friendly critters that don't mind being handled. We could offer families with little kiddies a fun day out. And have a few little merry-go-rounds and stuff."

"You'd need a huge capital injection to set up all those things," said Justin. "What puzzles me is how you've got into this situation. Forget about Albertelli for a moment. From what I've seen, it seems quite hard to make this place pay. Costs seem so high and the returns don't look good. How have you managed to survive since 1745?"

"The rents paid for nearly everything, so we didn't need to lift a finger," said Lucrezia. "We had ten tenants, all paying us rent. We kept ten hectares of gardens and paddocks for our horses, and all the rest was farmed by the tenants. We didn't need to think about making farming pay. And we had other sources of income as well. Commodity trading and soldiering are as old as the hills. And most people don't expect their house to make them a living."

"So, what's gone wrong?"

"Farming doesn't pay well any longer," said Iska. "It's the cheap food coming in from abroad, and the supermarkets paring the prices to the bone. Half the tenants have thrown in the towel and handed the land back to us. We've only five tennants left now, so we're five rents down and we've got all this land we're not sure what to do with."

"Phew!" breathed Justin. "How long has it been like this?"

"Ten years ago everything looked fine." said Lucrezia. The tenants said they were getting better prices than ever. We didn't realise things were starting to go wrong. The tenants all seemed to have good personal reasons for giving up, such as ill health, or no children prepared to take the farm on, so we didn't give it much thought. We expected people would be queuing up to move in. When nobody applied to take it on, Iska offered to farm it and took on the farm hands who wanted to stay.

"To be honest, living here had been so undemanding for centuries that we'd lost interest – thought we ought to do something to help suffering humanity instead. When we decided to train as doctors and join Medecins Sans Frontieres we sort-of mothballed the estate, cut back, got rid of the flower gardens so we needed fewer outdoor staff. The House-keeper and Den acted as caretakers. I can't believe how irresponsible we've been. It's only now that someone is threatening to take it all away that I've realised how much we love it. And we're pretty sated now with dangerous adventures. Sorting out Monrosso suddenly seems the most exciting thing to do. We've got to save it somehow."

"Well, then, back to the hacking this afternoon," said Justin. "Maybe you could raise a loan to tide you over."

"I've tried already. We've had so many EC grants they won't consider any more, so I'm trying to make a case to the banks. They want to see the statements for the last three years and a forward business plan. The accounts Albertelli sent me look so fraudulent now I can't possibly use them. If you could somehow get me our real bank statements, at least I'd know the truth and whether we have a sporting chance of making the place pay."

"I'm doing my best, but I'm not getting anywhere. The pin number was easy, but the password could be absolutely anything. Has anybody any ideas? Does he have any kids? What's his wife's name? Dog's name?"

"No, no kids. Wife? Elenora. Dog? Bonzo," said Den.

"Bonzerosso!" said Iska. "He used it as an expletive as well."

"Bonzerosso! That would make a fun password. I'll give it a try this afternoon," said Justin.

"Bonzerosso, open the box!" Justin muttered through gritted teeth as he typed it in. If it didn't work it was back to more slog, reinventing his old password-cracking algorithms from very hazy memories. It could take a week. The banks had grown more skilled in self-defence since the old days, having been hacked far too often for their customers' peace of mind.

"Eureka! Done it!" he shouted.

Lucrezia came running up the stairs of the Lodge.

"You've found the account?"

"Think so. Certainly found an account. Have a look at this and see if it looks like the right one."

"Well, there's a payment from the co-op at the start of the month; there's the rent from the riding school; there's one from the deli in Menaggio; Eurobet! Eurobet, Eurobet. All going out, not coming in."

"Well, they would, wouldn't they?" scoffed Justin. "I heard somebody say the betting shop had been after him. He's certainly in the red, look! And look at the bank charges. Must be unauthorised overdraft charges. Well, you wouldn't want your name on an account like this, would you?"

"But what's he done with all the money from the EC grants? He hasn't paid Tekoverde. The drawer's full of bills for expensive Tekoverde machinery."

"I'll download all the statements for the last three years and email them to you so you can get to work. Then I'll see if I can find a savings account as well."

The current account went back only nine months. Justin printed off the nine statements in case Den accidentally wiped them off - it was Den's laptop.

"Thirsty work, hacking," he said as he dumped them on her desk and went off to the kitchen to put the saucepan on for tea. "You need a kettle," he grumbled, bringing in the mugs. "Tea tastes wrong with pan water." At least there was fresh milk now in the fridge. He'd taken the milk powder up to the refuge.

"Nobody sells kettles in Italy. You've only given me nine statements. I need three years' worth to get a loan."

"The account only goes back nine months, and nobody would give you a loan on that account, would they? He's been helping himself to the bank's money as well. Now we need to do two things: find his savings accounts, if he has any, and find the accounts he was using before nine months ago. They were presumably with your old bank, the Popolare. Wish me luck."

Maybe the tea helped. It was real Yorkshire tea, after all. By four he had tracked the transfer of the account back to the Popolare and got her another two years and three months' worth of statements. Now he could attack the problem from both ends. By six he had both the old and new savings accounts.

He studied the more recent savings account with excitement. It seemed to be still in the black, though the figures had shrunk considerably in the last few months. Eighty-six thousand. Not much, considering the money coming into the earlier statements, but not exactly chicken feed. Over a hundred thousand had been sent off somewhere earlier that month. Had Albertelli opened other accounts? Maybe this would soon disappear as well. Could he nail it down? After all, it belonged to the Marquis, surely, not to the agent. Once he had managed to work out how to grab it he hesitated. He didn't know Lucrezia's bank account numbers. If he wasted time asking, this last chunk of money might disappear like melting snow. On an impulse he typed in the number of one of his own secret Swiss accounts – an almost empty one. If he was not too late and it arrived there he should keep it as a nice surprise - safe from the bankruptcy receiver who would have seized it. With their Rome house in ruins they could end up homeless and penniless.

"Justin's cracked it!" Lucrezia exulted, as they sat down to dinner. "He's located the bank accounts."

"Unfortunately there's nothing in them," he cautioned. "Worse than nothing. All overdrawn."

"Well, at least we know where we are," she sighed.

"And you can't be held liable for his overdrafts, since you'd agreed to take your names off the accounts. You can be thankful for that small mercy. I'll carry on studying the accounts, see if I can see where any of the money went to, but I guess it might all have gone to Eurobet."

"So," said Den, "If Chez can't pull any rabbits out of the hat we're sunk. Shame. Nice living here, on the whole. Beasts can be a challenge, but they're good for a laugh."

"Good fun chasing the drunks last night," said Iska. "Wolves are shaping up very well, don't you think? Without the mad one they're just pussy cats."

"We'll have to find them a zoo. They'll hate that, won't they? We've ruined things for them. They're too tame now to be left free to roam. They'd hang around near people and get shot in no time. Chez will be heart broken. If only we knew where he is. It's been ten days now. He usually gets things done in a few days. Maybe he's landed behind the dustbins, and they've locked him up somewhere. He might be trapped under the rubble again. We should go look for him, dig him out. Forget Monrosso. He's more important than any rotten country estate." Her eyes were swimming with tears.

"We do know where he is. He's working in the medical school in Raqqa," said Justin.

"You've heard from him? When, how?"

"No, of course not. He'd give the game away if he tried to communicate with us. He told me that's what he intended to do. Surely they'd be delighted to have him."

"I thought he masqueraded as an arms dealer to lure the fanatics to him," said Den. "They might never go near the medical school, so how is he going to find them?"

"That's what I thought," said Justin."It may take weeks."

"We don't have weeks," moaned Lucrezia. "A week tomorrow it will all be over. I wish Vetriano's would get back to me. They promised to fix me a meeting with the tax man."

"What good would that do?" asked Den. "What are you going to do, flutter your eyelashes or offer him a bribe?"

"Put off the evil day, that's all. If I can convince him the tax demand might not be correct he can let me pay only thirty per cent of the bill, then go to a tribunal. At least that would give us a few more weeks to find the money."

"Judging by what we've earned this week it'll take us about fifty years," growled Den. "Let's face it, we're stuffed."

199

"So," said Lucrezia,"are you going to scuttle off, leave the sinking ship?"

"Are you calling me a rat?"

"Well, are you?"

"I may be a rat but I'll wait till we've actually sunk, if that's all the same to you."

"Wouldn't miss the denouement for the world," grinned Iska.

"You actually think this is funny, do you?" Lucrezia exploded.

"Look, Krish, I know you think this is all our fault -"

"No, I don't. It's our own fault. We've been neglecting this place for years, so we've no right to blame you. We took on estate managers in the past, now and then. We were lucky they were all honest men - at least they didn't bankrupt us. If we can somehow weather this I swear I'll run this place myself and make it pay."

"So," said Justin, "let's get the shop fixed up. The tea party doesn't start till four tomorrow, so we've time to do another coat of whitewash if we start before breakfast. Who doesn't mind getting up early?"

CHAPTER 23

BAGS

"Bags!"

"Argh! Ooh! What's the matter?" groaned Justin.

"We've no bags. Bags for the farm shop. How can we sell anything if we've no bags to put it in?"

"Crikey Moses, couldn't it wait till morning?"

"You can help me remember. I might have forgotten by morning. First thing tomorrow we have to get some bags."

"Okay, okay, now can we get some sleep?"

"You've had some sleep. Are you sure you want some more?" she purred.

How many arms and legs did she have? They seemed to be invading him from all directions, smooth and sinuous as boa constrictors. Oh, to hell with sleep!

"Come on, then, let's go find some bags," he said, tickling her awake.

"Argh! Ooh! What's the matter?" she groaned sleepily. "It's the middle of the night."

"No it isn't. Sun's up, birds are singing. We've time to fix the bags problem before breakfast. Race you down to the Lodge."

"Don't do racing. Breakfast in bed would be nicer."

"The lady said bags, first thing in the morning. Very important, so she said."

"Aooh!" she groaned. "What if you try to find some by the time I get down to the Lodge? I'll give you a good start."

None of the neighbours seemed to have turned on their Wifi hubs so early, so he left the computer struggling to find an active hub while he made two mugs of tea and scoffed half a packet of biscuits. Damn it! All that jogging down the drive had been negated by a few greedy scoffs.

To his surprise she did actually make an effort to lend a hand, bless her cotton socks! Here she was, a little bleary-eyed, before the tea had had time to go cold – and just when he had found something worth showing her: a distributor of plastic bags in Como.

"You've got a mobile, haven't you? If you rang and asked them very very nicely, who knows? They might just pull the stops out for us."

Just as he expected, she was very good at wheedling.

"No, honestly, how amazing! You must have second sight. Dark green, like Harrods in London? Yes, very smart. I'm sure the Marchesa will love them. And they say Monrosso House? You're wonderful, a knight in shining armour. Can we come and pick them up? Well, you see, we're trying to open the shop this afternoon. Yes, we've a party of local ladies coming, so I expect they might want to buy something. Yes, of course, I'll ring back in ten minutes."

She switched off the phone and turned to him, smiling from ear to ear. "You won't believe this, but their manager saw our notice board. They've printed 'Monrosso House' on a few bags and are planning to send them on spec, hoping we'll order loads more."

"What have you two been up to?" asked Den. "You look far too pleased with yourselves."

"Ha ha!" said Kerallyn, "we're not slouches. We've just organised some bags for the shop. They should arrive mid-afternoon today."

"Mid-afternoon? That's far too late. Can't you go and fetch them? Where are they?" asked Den.

"Como, but they won't let us. They're already on the van and they won't consider trying to locate them. We'll have to wait till they arrive."

"Sounds like a recipe for disaster to me," said Den. "Probably arrive just when everybody's gone home. We'd better ask all the staff if they can lay their hands on any used ones as a back-up."

"I've found a few plastic trays to put the fruit in," said Cook, bustling into the Morning Room, "but you'll surely need lots more." She put six trays on the breakfast table.

"Ah, ye gods and little fishes!" said Den. "We're kidding ourselves, thinking we can open the shop today. We need all kinds of bags and containers, and we ought to have had the inspectors in as well."

"Look, I told you, we had the inspectors in when we started packaging our own meat and cheese," said Iska. "They came to check the new wine vats and the bottling plant as well. I've checked the regulations and we should be okay. We've got a permit for the market stall, and we'll be selling all the same stuff."

"Mmm, all sounds well and good, but they always seem to find some red tape to strangle you with. Maybe they expect a back-hander."

"We'll go foraging this morning, then, shall we? Or is there anything else you need us to do?" asked Justin.

"Can't think of anything. Maria and the girls are helping Cook make the cakes and sandwiches," said Den.

"The lads are going to put the tables in the Orangery once the whitewash is dry and the men have tied the climbers back up," said Iska. "Once we've got the chiller cabinets in place the wives will bring all the produce up and arrange it. We should have everything in place by lunchtime."

203

Maybe they should have gone to Como. It was a self-indulgence to drive north in Iska's red Spider, but the picturesque little lakeside towns offered poor pickings. Nobody sold plastic shopping bags, and surprisingly few used them for their shops. They managed to wheedle about a couple of dozen altogether from all the shops they tried. It was amazingly difficult to find any plastic trays as well. Even small plastic bags were hard to locate. Everybody bought them from supermarkets, but there seemed to be no supermarkets around. They bought a few paper bowls from a fancy goods shop. They might hold a few soft fruits, failing all else.

"We did our best," he said. "Sorry. We should have gone to Como."

"We'd better ask around the staff again, for plastic trays and little bags this time," said Lucrezia, as they hurriedly chomped through lunch. "Thank you for trying. Sorry you had to do it. I wish we could give you a better holiday - "

"I'm finding the whole thing quite exciting. More fun than the Himalayas. A new challenge every day."

"Exciting? Huh!" said Iska.

"That's a novel way of putting it," said Den.

"So, now, what else do we need to think about?" asked Lucrezia. "They are expected to arrive about three, so first we take them on a tour of the House. I'll have to do it this time, as Maria and the girls will be doing the food. Are any of you free to help?"

"I am" said Justin and Kerallyn in unison.

"I'd better stay outside to keep an eye on the animals," said Iska.

"I'd better stay in the entrance hall to see all's well in here." said Den. "And I need to supervise the tea things going out. Maybe Maria should be in the Orangery to welcome the guests and check the food arriving."

"Well, everything seems to be covered, then. Shouldn't be a problem. Only afternoon tea for a few local ladies," said Iska.

"It's the farm shop that's worrying me," growled Den. "Crazy half-baked scheme," he added under his breath.

Iska gave him a scowl and stomped off outside.

"Oh, is that it?" the lady with the unbecoming hat asked stridently, when the group emerged through the basement. "Not very big, is it? We went to Versailles a few weeks ago and Blenheim Palace in England. They were something to shout about; not a bit like this."

"Monrosso has a similar claim to fame as Blenheim," said Justin. "Both palaces were a gift from their grateful countrymen to their greatest heroes. The difference in scale is just as it should be. England in the early seventeen hundreds had the biggest, most powerful empire the world has ever seen. John Churchill had just won a few great victories against England's most powerful enemy, France. A very wealthy grateful nation made him a Duke and gave him that magnificent palace.

"In those days Italy didn't even exist. It was just a few small weak states. The most powerful man around here was the Duke of Milan. He gave Cesare, the first Marquis, the title and the estate as a reward for his great victory over a crack Swiss army. And he ordered him to defend the people hereabouts from any more trouble from the Swiss. As far as I know there's been no more trouble from the Swiss, has there? I hope the people who live around here realise what a good job all the Marquises have done for them all these years." Put that in your pipe and smoke it, he growled to himself. Bet she lives in a scruffy little dump!

"I was hoping to see a picture of the Marquis. Is there one somewhere?"

205

"No, that's another thing," grumbled she of the ugly hat. "Most stately homes have lots of pictures of ancestors. I haven't seen any."

"That's right," said Kerallyn. "The Marquises have always been men of action who hated dressing up and posing for painters. They don't like to show off."

"Must be ugly. What does the current Marquis look like? Number Seven, isn't he? Is he ugly?"

"His two brothers are both around here somewhere. They all three look quite similar. One manages the farm and the other the House. They look quite handsome to me."

"Versailles is full of gorgeous paintings of the kings and their ladies," asserted The Hat.

"Yes," said Justin, "and just think where all that showing off and extravagance got them. It made their people so furious they guillotined the lot of them, didn't they?" Game, set, match, thought Justin.

"Do you feel ready for tea?" asked Lucrezia. "We have just set up a little cafe in the Orangery, and you will be our first ever guests there. It would be a shame to have tea indoors on such a lovely day. Please follow me."

"Oh, very nice and colourful," breathed Kerallyn.

There were bright paper tablecloths on the old bar tables and the crockery was in half a dozen different styles and colours. Outside on the terrace were two more tables under yellow parasols.

"We had to search high and low for the tea cups," whispered Lucrezia. "Tea isn't drunk very often in Italy."

"Do you have coffee instead of tea?" asked Big Hat. "Can't stand tea. Can't imagine why the English like it."

Kerallyn scampered back to the kitchen and called for a good supply of coffee. As she anticipated, a dozen of the ladies followed Big Hat's lead.

They seemed to enjoy the dainty sandwiches and delicious little cakes. The Orangery turned out to have a very pleasant, ringing acoustic, perfect for the speeches. Finally the ladies trooped out and headed for the shop.

"Oh, no fancy goods, just food," moaned Big Hat. "No good to me." She stomped back to the Orangery, dumped herself onto a chair and looked impatiently at her watch.

A plump woman in a flowery dress poked around the chiller cabinets enthusiastically. "Oooh, just look at this lot," she gloated. "Have you got a basket?"

"Basket? Oh dear!" whispered Lucrezia. She hurried over to the corner of the garden, picked up a flower trug and gave it a good shake.

The woman gave it a surprised look, then proceeded to shovel things into it excitedly.

"How much are these strawberries?"

"Three Euros the big boxes and two the small ones."

"And the middle sized ones?"

"Erm, two. Buy one of those and get a bargain," said Justin. "Quick, before they all sell out."

There was no till, just a big mixing bowl for the money, a receipt book and a pen. For the first ten minutes they were all turning out their pockets and appealing to their customers for change. From then on it was downhill all the way. They soon ran out of new bags, all with other people's logos stamped on them, and had to resort to scruffy-looking used ones. When the plastic trays ran out the soft fruits had to go into small plastic freezer bags. They'd be squashed by the time the customers got home.

Cook came bustling in with dark green plastic bags.

"These have just arrived. Thought you'd need them."

"Oh, thank Goodness!" breathed Kerallyn. "Oh no! I

was sure he said two dozen or so. There's only two. I must have mistaken due for dozen. This doesn't seem to be our lucky day."

The fat lady was the last to sing. She proffered a credit card.

"I'm so sorry, but we can't accept credit cards," said Lucrezia.

"What?" puffed the woman, reddening. "There's nothing wrong with my card. Plenty of money in the account. What do you mean, you can't accept it?"

"Oh, my Goodness, I don't mean to suggest – It's just that we don't have any means of dealing with cards of any sort. We'd need a card reader and we don't have one. We can't, you see, because we don't have a phone line here yet. We'll get one soon, I hope."

"You've got a phone right there," accused Big Hat, who had come to try to chivvy everyone off to the bus.

"That's just an in-house intercom. It's not connected to an external line. Too far from the main phone lines, you see. Would you like us to deliver the things to you? Then you could give us a cheque."

"My bank doesn't deal with cheques. Everybody uses cards these days."

"You can't run a business this way," shouted Big Hat. "It's just ridiculous."

The fat lady threw the trug on the floor and stormed off.

"Oh dear!" sighed Lucrezia.

"You go sit down," said Justin. "We'll clear up."

CHAPTER 24

BAND? WHAT BAND?

"You don't know the worst of it," said Lucrezia, over dinner. "That awful woman is refusing to pay us. Insists she can only pay for the group by card, not cheque. Seems we may have done thirty-two teas for nothing."

"Let's think," said Justin."What about the riding school? They might use a card reader. Could you trust them to have it paid to them, then give you a cheque or cash?"

"Brilliant idea. I'll call in tomorrow morning. And I've had one bit of really good news: Vetrianos emailed at last to say they've got me a slot to see the tax man on Monday. I'll have to go to Rome on Sunday afternoon."

"That's nice," said Justin. "We can all go together, then. I have to start work on Monday and so has Kerallyn."

"No, you're surely not leaving, are you?" There was a surprised groan from everyone around the table. Yes, it was only polite to feign sorrow at the loss of your guests, but this seemed quite genuine. Had he really ingratiated himself into the lives of these unique beings? Wow!

"I'll have to find a hotel -"

"No, you won't. You have a home in Rome, remember? I told you and Cesare my penthouse was yours to use any time you like. The same goes for you, Den, Iska, Kerallyn. I've two huge comfortable sofas you'd enjoy sleeping on, so we could all live there at once, if ness."

"You're very kind. We really do appreciate that. Now, about tomorrow night. The Silver Wedding people will be arriving from about seven. There's about seventy of them so we can't afford to give them an excuse not to pay. I'll ring them first thing tomorrow and ask them to confirm they'll give me a cheque on arrival. Be on the safe side.

No livery or fuss this time: they just want to eat and dance. So, everybody in simple black, men just black shirts and pants. Anybody hasn't got a black shirt?"

"I'll check with the lads straight away. They can help the girls collect up dirty plates and glasses and things without any training, can't they? It's just a buffet," said Iska. "We three men can manage the buffet."

"Well, they were fine last time," said Den, "so I don't see any problem there. I've ordered loads of cheap glasses and more cutlery and plates from the hire shop. And fifty stackable chairs. The men will have to bring the furniture back from the Orangery. Got lots of smart paper napkins. Cook and Maria have already started the food, and the wives are coming up at six tomorrow, as usual. Everything should be okay."

"If you've nothing else for me to do I'll go down to the Lodge and book our train tickets in the morning," said Justin, "and look up card readers and tills. I've a feeling you can get a card reader that works with a mobile. At least you could work it from the Lodge."

"So, this will be our largest number of customers yet. Loos! Can anyone remember how many the Inspector said we had to provide?"

"A lot less than I expected, only about three of each for over a hundred people. With the Portaloos we don't need to let anyone go upstairs," said Den.

"Won't the women be too scared to go outside in case the wolves get them?" asked Justin. "

"Mmm, well, we'd better make them men only. If the men are scared they won't dare to admit it. I've got men on duty with tranquilizer guns, just in case," said Iska. "The haha seems to be keeping the rest of the animals away."

So, everything sounds to be under control," said Lucrezia, "thank goodness."

Famous last words!

"When is the band due to arrive?"

"I'm sorry?" queried Lucrezia. "The band? I don't know. I was just going to ask you."

"Me? How would I know?" asked Mrs Capone. "It was your staff, I presume, who made the arrangements."

"My staff? No. Nobody asked us to do that. The last party brought their own band, so we assumed - "

"Brought their own - Well, what on earth made you imagine we had our own band?"

"No, I mean that the Castle Hotel had arranged it. Haven't they arranged one for you?"

"Carlo, did you talk to the hotel about booking a band?"

"Me? Why would I do that? You made all the arrangements. So, what's happened to the band? It's about time they got started."

"There is no band. These crazy people imagined we'd bring our own."

"What! Are you telling me we book a dance and get no band to dance to? What kind of dopes - ? We'll be the talk of the town, sneered at from here to Naples. This is sheer negligence. We'll have you in court – we'll - "

"Excuse me, My Lady," said Iska. "The musicians will start in about fifteen minutes. I'll just go fetch them in."

"Will they? Well, there you are," smiled Lucrezia. "Please excuse me." She hurried after Iska who was propelling Luigi very fast towards the kitchen.

"Luigi, you know the hit parade. I hear you singing all the time. You said you had a keyboard. Is it working?"

"Was last week. Gianni and me took it round to Kit and Sofia's and we had a right singalong."

"Right, you're having another right singalong tonight.

Get down home quick as lightning. Take the Jeep, it's here in the yard. I want you back here with that keyboard before you've even gone. Now, quick, get going. Right, everybody, I assume we can all play something, so let's see what we can find in the Music Room."

"What?" exclaimed Den. "I don't know any pop music. And with no scores and no time to rehearse - "

"Have you any better ideas?" demanded Iska.

Den threw up his hands. "You win. What instruments can you play, Justin?"

"Comb and paper," said Justin sheepishly. Still, with twelve thousand years to kill, they'd surely had plenty of time to learn pretty well anything, so no need to beat himself up about it.

"Cook, can you find him a bowl or a bucket and a couple of wooden spoons? Yes, that's fine. Here's your drum kit, Justin. Enjoy. Maria, can you find us a long electric lead and plug it in a socket on the dais in the ballroom? Now, Music Room, everybody."

By the time Luigi rushed in with the keyboard, Lucrezia was tuning up a violin, Den was strumming a double-bass and Kerallyn was trying out a few trills on a clarinet.

Iska put down his saxophone and plugged in the keyboard. "Now, Luigi, does this thing have any built-in tunes it can play automatically? Yes? Right, now you just set it going on pop tunes they can dance to. Don't let it play anything solemn like 'Ave Maria,' for Goodness' sake. We'll try to join in as soon as we've got a grip on the tunes."

"Are you going to announce us? Do we need a name?" asked Justin.

"Not bloody likely! Just sneak up on them, get a bit of practice before it occurs to them to listen," Den muttered.

There were seven pop tunes, some more and some less suitable to dance to, and the dancers did a fair bit of grumbling at first. Whether they got tired of grumbling or the band improved with practice there was no way of telling, but once he got the hang of the spoons and bucket, Justin began to enjoy himself. Of all the lunatic ways to spend a Saturday evening!

"Mr. Chase, listen to this!" Cook hissed into his ear. She put down a big metal mixing bowl and ratatatted a big spoon from side to side inside it.

"Brilliant!" he said, taking the spoon from her. It was a bit tricky playing both the dish and the bucket at the same time, but he got the knack in the end. "I play the dish and bucket." Much more interesting than comb and paper.

"What now, Mr Romanov?"

"Play them again, Sam."

"Who?" asked Luigi.

"Just play them again. Go on!" Iska laughed, then gave a virtuoso performance on his saxophone, prising a burst of applause from some of the listeners.

Trust him to grab any chance to show off, thought Justin.

"Your turn, Kerallyn," said Den.

Very impressive, thought Justin. She could certainly play like a pro. If I really am immortal there'll be loads of time for music. Can't be bad, can it? He pounded away on his dish and bucket in increasingly complex rhythms.

The others stopped and let him do a solo, laughing as he acknowledged the applause.

"Now what, Mr Romanov? I've played them all twice over."

"Right, time for you and Gianni to play us something up-to-date. Don't worry about mistakes. Once we get the tune in our heads we'll try to drown you out. Come on."

Blushing furiously, both lads did their best, but it was soon clear they needed drowning.

"You're so good at singing," said Lucrezia,"why don't you both do that and I'll try to play this thing."

"Crikey Moses!" muttered Justin."They're bad enough to curdle milk!"

"I've heard far worse in the clubs," said Iska. "Typical of your average boy band. And just look at those girls."

A gaggle of girls were giggling and clutching each other, and fluttering their eyelashes at the lads.

"We found them first," said Judy. "Come on, Heidi, let's show those girls they haven't a chance. We've finished clearing away, so we're off duty now."

Now there was a pop quartet, wailing off-key and waggling their hips.

"Ye gods and little fishes!" groaned Den. "How much worse can it get?"

"Awooo! Awooo! Awooooooo!"

Outside the huge open windows the wolves threw back their heads and joined in the chorus.

CHAPTER 25

GOODY GOOD!

"Are you sure this is the right address?"

"Let me see the paper again," said the taxi driver. "Yes, definitely: it's the right address."

"Looks a bit - "

"Yes," said the taxi driver, "best not to walk around here after dark. Know what I mean?"

"Stacey warned me about that. But she was okay. So, goodbye, you two. Hope to see you again soon."

"We'll come in with you and make sure everything's alright," said Justin.

Kerallyn unlocked the door and led them up the stairs.

"Stacey said it's on the top floor. Good exercise, isn't it? Oh, listen. Seems to be somebody in here."

It sounded like a TV set at high volume, and a clatter of pots and pans. Justin had to knock hard to attract attention.

"Hello?" The tousle-headed youth looked puzzled.

"Sorry to disturb you. I'm looking for flat seven. I must have the wrong flat."

"This is flat seven." He looked at the paper she held out to him. "What? You've rented this flat from three weeks ago? You can't have. I've been here a week already and I've got three more months left on my lease. We'd better go see the landlord. Let him sort it out. He lives downstairs."

Goody goody good, thought Justin, listening with growing delight as the landlord explained that, as he had not heard from her for three weeks, he had assumed she wasn't coming, and relet the flat.

"But I did explain that I wasn't moving in when Stacey moved out because we were having a holiday in Como."

"Did you? I don't remember that. Oh dear! Well, I'll have to ask around and see if I can find you something else. I'd offer to put you up for the night, but I don't have space for all three of you."

"No problem," smiled Justin. "Mrs Wilson can stay with me, and I can find her a nice flat. Did you pay a deposit? Yes? Well, if you could refund her deposit, we'll go and leave you in peace."

Oh, what luck! Justin gloated silently as the taxi carried them away from that rather insalubrious district towards his ancient palace. If he couldn't persuade her to move into his penthouse he could surely rearrange existing bookings and let her have one of his six tourist flats.

"Oh, this area looks much more interesting," she said. "Shops right here, in the same building. So different from The States."

"Hold your hand just here," he told Lucrezia, when they stood in front of the intimidating black doors of his penthouse.

Lucrezia obeyed, but nothing happened. "Now you," she said. "This should be interesting."

Justin swallowed hard and slowly held his palm in front of the door mechanism. "Oh Grief!" he moaned. "Can't even get into my own home now."

"A nice new special gift for you," she said. "Great to be immortal, isn't it?"

"What's supposed to happen?" asked Kerallyn.

"The door's supposed to read your palm and open, of course. If you're a poor mis-shapen Immortal you have to press this secret button behind the door frame, here," said Lucrezia.

"This is just the sort of thing that's worrying me, just

what I'm tryng to research. If I can get enough time alone with the right machinery - Oh, what a lovely flat! You lucky man!"

"You like it? Well, why don't you unpack your suitcase and hang your frocks in here? Lucrezia, you still have a few frocks in the guest room, so are you happy to move back in there?"

"Where's your room?" asked Kerallyn.

"He sleeps on the sofa. Absolutely insists on it, don't you, Justin," Lucrezia smirked.

"Absolutely," he grinned. "Now, must be tea time."

In fact it was a little late for tea, but he made some anyway, as a kind of welcome home.

"Now, dinner. Do we cook - ?"

"What about that nice little restaurant again? We haven't been there for a couple of weeks," said Lucrezia.

"A couple of weeks? Is that all it's been? So much has happened it doesn't seem possible."

"Two weeks ago we thought Monrosso was Paradise, and now it's getting more like Hell."

"It is Paradise. It's just that serpent Albertelli. But we're not going to give in, are we?" Justin asserted. "We'll find a way to sort things out. You're all so clever you're bound to think of something, and have you forgotten Cesare? He's our fairy godfather. Bound to save the day."

"Of course I haven't forgotten Cesare. I'm worried sick about him, but I can't just sit down and cry. I have to be worthy of him, don't I? Keep everything afloat till he gets back."

CHAPTER 26

BACK TO THE GRINDSTONE

Walking the floor, pressing the flesh – what every caring boss should find time to do at any appropriate juncture."How was the cruise?"Did the wedding go well?" "Did the operation fix the problem?" "How has your grandma settled into the nursing home?" "What a tan! You look fantastic." "Welcome to 'Just in Case'. If you need any help to settle in, just give me a shout." By the time he'd done the rounds it was coffee break, thank God! After a word with anyone he'd missed, he took refuge in his office and sought solace in his mug of tea.

Marcella had put the sales figures for the original killer zombi game on his desk. The trend was clear to see. After three years of heading for the stratosphere the graph had turned resolutely downhill. Was there any country left unsaturated? China, India, Brazil, Indonesia, the EU, all the big markets showed the last few drops of profit being steadily wrung out. Surely this cash cow was running dry. Definitely time to let it go. Lucky there was a buyer keen to snap it up. Fruitos Beers wanted to give the famous game away as a promotion. No point in trying to sell it after that, but it would soon be hard to sell in any case. Yes, take it away, Fruitos Beers. Just offer me a ton of money and it's yours. When are you coming? Two-thirty today. Great.

Lucrezia was meeting the tax man at two-thirty. Good luck, Lucrezia. And Kerallyn had an appointment with somebody big at the UN Food and Ag. Good luck, Kerallyn.

"Paolo is away on business next weekend," smirked Marcella.

"Really? Mmm, can't wait," he murmured, giving her a

218

knowing look. Better get the auditors in quick, he thought.   After the depredations of Albertelli it seemed you'd be an idiot to trust anybody – least of all your finance director.   And what was he going to do about the gorgeous redhead in his bed if Marcella tried to move in for a day or two as usual? Everyone should have such problems.

Why did it all seem such a drag? Let's get the day over and go home, back to helping Monrosso with its fascinating real-life problems. Oh, the new computer game. He hurried up to the Think Tank, impatiently waved away their clamour of questions and info, and launched into his idea for the contest between the estate owners and the bailiffs.

After the longest silence he had ever heard them stage, they all shouted at once. "Yes, but - ." "Are you serious?" "You can't be serious!" "Hey, but listen, the bailiffs are really cyborgs - " "Don't tell me, the owners are really zombies?"

"Look," said Justin, "This is something we could aim at a completely different market. Loads of older people, even pensioners, are into computers these days. They've been emailing their grandkids forever, must be sick of reading the same online news rags day in, day out. They must have done crosswords and Shanghai and Freecell till they're blue in the face. Can you imagine them playing with zombies and cyborgs, mad car chases, exploding tanks and stuff?"

"My grandad plays 'Grand Theft Auto' till two in the morning."

"And what does your Grandma do?"

"She's dead."

"Well, there you are. But I'll bet it wasn't through playing Grand Auto."

"No, she dropped dead with shock when she won first prize at Bingo."

"Exactly! Just the point I'm trying to make. You're not exactly thick, so I bet your Grandma was really pretty bright. Must have deserved something a bit more taxing than the Bingo. This game would be for her, and all her pals. She's had a lifetime of experience to call on to try to solve some real-life problems. If the bailiffs won she might shed a few tears. If she saved the estate she'd be cock-a-hoop for days, wouldn't she?"

"Well, you may have a point, but how do we do it? Don't know the first thing about estates and bailiffs," said Gianni.

"Well, I do, so you can just ask me - or Google. I want you to mull it over and do a few simple scenarios. You finished all the stuff we were doing before the holiday. Most of it seems to be working well, so Gianni and Freddo can do the snagging on the 'Zombie Bedlam' they're complaining about and the rest of you can see what you can come up with for the grannies."

Bedlam broke out, of course, so he made a quick exit and went back to the Zombie Original figures. Amazing, this appetite for mayhem. What did it say about human nature? Well, just look at the newspaper headlines. It was certainly not just fiction: all over the world they were doing it for real.

And somewhere, behind those chilling headlines, Cesare was really fighting his way through it – almost certainly unarmed. Of course Lucrezia had insisted on going to the ruins of the Borgia House on the way home, and this morning too, but there was no sign of him there. Obviously he really was having trouble finding the rats, with no armaments to offer as bait. Hmm! You'll need far more than good luck, Cesare, but be quick, or the bailiffs are going to win hands down!

CHAPTER 27

FANCY FINANCE

There was no need to ask how things went. They were dancing to the radio, wearing identical Marilyn Monroe white frocks, when he managed to get into his flat, having absent-mindedly stood for a while holding his useless palm in front of the door mechanism.

"Yes?" he demanded.

"Yes! Yes! Yes! I can go to the tribunal. Everybody says it will be ages before they can fit me in, so we're saved - for the moment."

"What about the thirty percent?"

"That's a hundred and eighty two thousand. I've got that in my private account. They want it by the end of the week. Thank Heavens! Thank Heavens! I've talked to Medici's as well and they've fixed me a meeting with the insurers and the Heritage people in the B House tomorrow afternoon. Would you mind if I stayed another night?"

"Look, how many times do I have to tell you that this is your place to stay whenever you want to? No need to ask, just come. You know how to get in. Move some more things here so it feels more like your home."

"Well, you really are very sweet. I really do appreciate all you've done for us. You're a real friend in adversity. We Immortals need to stick together, don't we? And Kerall has good news as well."

"Yes?" said Justin.

"They've awarded me the Stoniface Grant. They'll pay me twenty thousand for my living expenses for the six months and another ten for research expenses. They're going to pay me to research our genome. Isn't that nice."

"What? You've told them you're immortal?"

"No way, Jose! I've really been commissioned to look for genes coding for susceptibility to foot and mouth disease in sheep. That's my day job. Surely they won't begrudge me using the machines to view my own genome out of hours – if they catch me at it, that is. If they do catch me, what can they do? Put me in jail?"

"And meanwhile, back at the ranch?" asked Justin. "Has anybody been in touch?"

"Yes, Maria Lepanto says we've three coach parties coming for tea this week. They're all foreigners on package tours, and she's made it clear the tour couriers have to bring us cheques or cash."

"If they're staying in hotels they won't want any farm produce, will they?" said Justin. "They won't patronise our shop."

"The health inspector rang this morning to tell us to close it down till he's had a chance to inspect it."

"That horrible woman in the hat again?"

"She's apparently his mother-in law."

"Poor sod!" laughed Justin. "We washed most of the spiders out of the roof and hosed the manure off the floor, so what's he grumbling about?"

"Maria and Den are determined to sell these trippers something, so the wives and the girls are all working their socks off making organic sweets down in the Dairy, and trying not to eat too many."

"Can you buy organic sugar?"

"We grow our own sugar beet and butter and cream and fruits. We can make fudge and toffees and boiled sweets and chocolates and - "

"You can't grow chocolate in Italy. Even I know that," said Justin.

"We have three huge carob trees. Tastes so much like

chocolate most people wouldn't guess."

"What have we got at the weekend? Silver weddings?"

"Nothing. The Castle Hotel is opening at the weekend."

"Oh, no!" squawked Justin. Was this the end for him? They wouldn't need him any more to play waiter or source cast-off junk.

"But," the other two laughed in unison and spun around to show off their new frocks, "we've got a couple of gigs."

"A couple of gigs?"

"The youngsters are performing in Menaggio in a new little club and we're doing a Gershwin evening in the Palace Hotel in Como. Do you know anybody with a drum kit? You'd better borrow it and practise hard."

"I think I need a drink," spluttered Justin. "What can I get you two?"

His heart was thumping hard, his hands made the glasses tinkle, and he had to wipe up uncharacteristic spills off the counter top. They couldn't be serious. It was just their little joke. But if they were only joking he'd be left out, wouldn't he? These two gorgeous, fascinating red-heads would go home and forget him. Monrosso would become nothing more than a memory. His future would be nothing but ever more infantile video games and boring sales figures. And what if he really had become immortal? He'd have to cope with the whole shebang all on his own.

"So, tell me about this Gershwin gig," he said, fighting for self-control.

"Saturday night. Open air concert in the grounds of the Palace Hotel. The band they booked has had a big falling out, so they've cried off. Somebody who came to our Silver Wedding is at the conference there this week and recommended us. Can't afford to turn this down. Need another string to our bow."

223

"An evening playing Gershwin is just a self-indulgence, isn't it? I don't need paying," said Kerallyn.

"Well, then, you can put your fee in the Save Monrosso fund," laughed Lucrezia.

"I was going to anyway. All for the common weal. Shall we take our drinks out onto this gorgeous terrace?"

Justin hauled open the huge patio doors for them, then turned back to the telephone.

"Freddo? Hello? Yes, it's me, Justin Chase. Look, sorry to disturb your evening. Didn't you once play drums in a band? At Uni, yes. Quite a while ago, yes. Have you still got them? You have? Thank God! Look, are you absolutely up to your eyes this week? Got a lot on in your spare time? No? Sure? Well, look, I need help. Say no if you hate the idea – of course I won't hold it against you - but could you spare the time to give me a few lessons this week? Yes, just this week. You sure? Thank God for that! I'm in a nasty fix. Hope you can save my bacon. Have you any time tonight? You have? Yes, in about two hours? I'll be round. Can't thank you enough. Chow."

"I expect you two have a lot of history to catch up on," he said, as he joined them on the terrace. "I have to go see a man about a dog, so if you can stand an early dinner, then you can have an all-girls-together chin wag, can't you."

"So, how did you get on with the Heritage people," he asked, as he dumped his briefcase in his bedroom next afternoon.

"They're going to think about it. That means nothing will happen for ages," Lucrezia sighed. "And I've just had some terrible news that will definitely sink us."

"Yes?" asked Justin.

"Tekoverde are taking us to court next week. We owe

224

them over half a million for all that new machinery. I've only enough for the thirty per cent tax, and I have to pay that right now. I can't see any daylight anywhere."

"Did you try raising a loan?"

"Justin, you've seen our Monrosso bank accounts. Would you grant us a loan if you were a finance company?"

He shrugged and sighed. Too true. "Look, I'm selling an old game very soon, just waiting for them to come up with a decent offer. Then I could put that in the kitty."

"No!" she said vehemently. "We can't let you risk your business. You know you shouldn't let sentiment get in the way of common sense. Chez would never agree to put your business at risk. He'd say if we can't make it work ourselves we don't deserve to keep it."

"It wasn't your fault; Albertelli cheated you; it was daylight robbery."

"It was our fault for neglecting Monrosso. We deserve to lose it."

"Your Rome bank might give you a loan," said Kerallyn. "You still have a lot of money in there, so you'll look like a safe bet to them. Get the loan first, then pay the tax bill."

"If I try for a business loan they'll surely want to see our records. I bet if I said I wanted to buy something ridiculous like a Lamborghini they'd say, 'Yes, of course, My Lady'."

Which is exactly what they did. On Wednesday Lucrezia used all her considerable charm on the phone to the Managing Director of Tekoverde and he agreed to accept an immediate payment of a hundred thousand Euros and a solicitor's letter promising to pay the other four hundred thousand or so within a month.

Everyone let out a huge sigh of relief. The first day of

reckoning was now four weeks away. The two thousand Euro fee for their Gershwin concert was not going to make a tiny dent in the remaining million Euros' worth of debt, but it felt like a step in the right direction. Better to work hard at something than sit chewing their nails to the quick.

"Iska and Luigi have hired you a drum kit for Saturday. Luigi and Gianni have spent all their overtime pay on one for themselves; think they'll be pop stars by Christmas."

"Think big," laughed Justin. "We're even worse. Do you know how many players the big Gershwin works are scored for?   I've looked it up."

"Mmm, big jazz band, say twenty five?"

"Don't forget the full symphony orchestra as well," said Kerallyn. "But have you heard of the 'Reduced Shakespeare Company'? They do daft things like 'The Complete Works of William Shakespeare, abridged' in two hours. There's only a handful of actors. They've been playing to full houses right around the world for twenty years."

CHAPTER 28

ONE STEP FORWARD, TWO STEPS BACK

On Friday afternoon they all helped themselves to a POET'S weekend, In other words: piss off early, tomorrow's Saturday. They caught a mid afternoon Freccia Rosso, and made it to Monrosso for a late dinner.

Lucrezia told them all about the delays of execution she had gained from the tax man and Tekoverde, listened to all their accounts of selling lots of new sweeties, and praised everybody for their hard work and dedication.

That dedication meant lots of overtime money for the staff, thought Justin. How much would be left in the kitty after that? All these little schemes were like shooting peas at a rhinoceros - no impact at all. Monrosso seemed to be holed below the waterline and they were trying to bail it out with tea cups.

After dinner Lucrezia asked him to walk down to the Lodge with her, just to see if there was anything in the post that she should know about. There certainly was: a pile of letters, begging, pleading, demanding that their outstanding bills should be paid. A few hundred here and there, it must add up to a substantial amount of money.

Lucrezia's eyes were full of tears as she handed him a handwritten letter. The old lady was about to close down her little shop. It was making too little money to pay her rent, and it was unpaid bills, many sent to Monrosso, that were making her insolvent.

"All these poor people," she whispered. "We thought we were saving lives out in the Middle East, but we were ruining our poor neighbours here at home. I must pay these people. There's about six thousand left in my

227

Rome bank – minus one hundred thousand, of course."

"And there's the few thousand you were making every Friday, trading shares and futures - "

"That's all over. Well, I was using the money in the Rome account to work the market, of course, but now all that's gone, so how can I buy anything? Maybe a little interest might have been credited in our old Cayman Islands bank."

She keyed in the numbers of the account. "Oh, yes, more than ten thousand dollars, but it was transferred to our Rome account days ago, when we closed this account. We've already spent that. Oh good grief!" she gasped.

"What's the matter?" asked Justin.

"I should never have closed this Cayman Islands account. There's no account for the bounty to be paid into now, is there? If the paymasters sent it, the bank will have sent it back, won't they?"

Justin groaned. "You could be right," he conceded. "Could you open it again?"

"Top priority!" she gasped, "but they may demand to hear from Chez as well. After all, the money coming in will be addressed to him, not me. Oh, what a mess!"

"Is there anything I can do?" asked Justin.

She shook her head in frustration. "Thanks, Justin, but best just leave me to wrack my brains."

CHAPTER 29

LET'S FACE THE MUSIC

This is just a nightmare, like 'Alice in Wonderland', or 'The Hitchhiker's Guide', so don't panic, Justin told himself, as a couple of hundred fairly inebriated, demob-happy conference delegates shambled out and took their seats on the lawn.

There was only one mike and they gave it to Justin, as his singing voice was no match for the other Immortals, not in volume, for sure.

"Good evening, ladies and gentlemen," he said. "I trust you're all ready to go off to the far corners of the world tomorrow and spread the gospel of extracting maximum results from minimal input. Good show! So, it seems we should have signed on for your conference ourselves. You asked us to play 'An American in Paris' and 'Rhapsody in Blue'. Yes? Great, we love them both, so it will be a pleasure. You all know, of course, that they are both scored for about eighty musicians. The symphony orchestra doesn't seem to have turned up yet and our jazz band is somewhat depleted, but, hell, John Snow promised us free booze if we finished by ten-thirty, so let's get going! Or should we call the whole thing off?"

He hit the cymbal and they burst into song: "You say potato and I say potarto – let's call the whole thing off."

After a sympathetic burst of applause, they were off.

'Strike up the Band', without vocals, gave them all a chance to let rip. Justin pounded away at his array of drums, grinning at the memory of his dish and bucket. He'd listened to Gershwin CD's till he knew the whole lot by heart, didn't need any scores – which was as well, because he couldn't read music. Freddo had told him most pop groups couldn't read music either, and some of them earned millions, so why worry? Next came 'Nice Work if You can get it' - very apt.

'An American in Paris' was a severe test of ingenuity and dexterity. Justin wished he could be more use as each of the others contrived to play three or four instruments almost at once, dropping one and snatching up another at breakneck speed. When the audience erupted in sympathetic laughter they cheerfully guyed it up. Den put on pseudo-confused expressions and flailed around as if trying to work out which instrument to play – all without missing a beat.

Afterwards the ladies, in their Marilyn Monroe frocks, offered to drive home, so the three men could attempt to increase their fee by drinking the free bar dry. Lots of people came to slap them on their backs. Some even slipped them a few quid. Many tried to chat up the ladies and nearly got their faces slapped. The pair of them could make a fortune as Belles du Jour or de Nuit.

CHAPTER 30

OH DEAR!

If only he hadn't drunk so much! Justin groaned as Kerallyn pulled back the curtains and announced that it was a shame to waste such a lovely morning lounging about in bed. Her cheerful chattering hurt his ears, so, in self-defence, he dragged on shorts and 'T' shirt and announced he needed a run to clear his head.

Auch! Auch! Every step seemed to rattle his aching head. He slowed down to a walk. Why on earth had they done this to themselves? The other two had got pie-eyed just as quickly as he had, so Immortals – hang on a minute, he was immortal, too. So far it seemed little advantage – if you didn't count surviving a helicopter crash. Now, that, well, yes, that had to be an advantage, surely, to die and wake up feeling on top of the world.

The Refuge was ahead. He nearly always found himself heading that way. There were the bee-hives to check – so far unoccupied, but they said it was too late in the year to expect a swarm. And there was the remote possibility that he might hit the jackpot, find Cesare yawning and stretching there, feeling as fit as a lop. The bounty had always appeared beforehand in the Cayman Island bank when he told his paymasters he was all lined up ready to do the deed. As he had said, you can't undo a contract killing if the customer refuses to pay up afterwards. But what if the bank account was already closed, and the bank had sent the money back? The thought sent a cold chill down his spine. Could they be coaxed to send it again, or would they just conveniently forget their assassin? He would have no redress.

Hang on, there was somebody there this morning, under that self-same tree where they had all three woken up that morning about three weeks before.

"Cesare! Cesare! Hello there!" Headache forgotten, he broke into a run.

It was definitely Cesare. No ordinary man would be lying there naked, showing off a body like a greek god. With that huge beard he'd be a perfect model for Zeus or Neptune, a formidable figure in his absolute prime, a man with the awe-inspiring stature of a god.

Panting a little, he stared down at the figure. Cesare had his arm across his face. There were deep furrows above his nose and his mouth was a very thin line. His whole body seemed tight as a bowstring.

"Cesare, are you okay?"

The figure tensed even more, and held its breath, as if expecting a blow.

"Cesare! Hello! Wake up! It's me, Justin."

"What?" He sounded bemused. "Justin? Where are you?"

"Here, right next to you." He reached out and poked his shoulder.

Cesare recoiled as if he'd been stung.

"It's okay, Cesare. It's just me, Justin. Look at me. Move your arm and look at me."

"Can't see you," he groaned.

"Course you can't. You've got your arm over your eyes. Move it away."

"No, can't. Hurts too much. Can't see you anyway. Got no eyes."

"What on earth are you on about? You must be having nightmares. Wake up, wake up!"

Cesare groaned and pushed his face further into the crook of his elbow.

"You've got bright blue eyes," said Justin. "What are you on about?"

"They've gouged them out. I've got no eyes."

"My God, who did that? When did that happen?"

"Just now. Didn't you see them?"

Hmm, thought Justin. What do I do about this? He's taken leave of his senses, talking utter nonsense. I can see for half a kilometre at least and there's nothing and nobody about. "The wolves did that?"

"No, no, the fanatics. Keep away from me, or they'll set on you as well."

"Where are we, Cesare? I don't know this place." he tried cautiously.

"Near the Bagdad Gate, the Al-Hal market. Get away, quick, before the Hisbah see you! Go in the market. Hide in the crowd."

"Market? What town is this?" asked Justin.

"Raqqa. What in Heaven's name are you doing in Raqqa? Go hide, quick."

"I've come to fetch you home. Promised I would, if you got stuck, didn't I? What time of day is it here?"

"Sunset, of course. Can't you see the lovely sunset? Last thing I saw."

"That was yesterday evening. You've died since then. You've resurrected. You're in perfect condition now, like you always are."

"No, I can't see. They've gauged my eyes out."

"Cesare, you're having nightmares. Just calm down and listen to me. You're not in Raqqa any more. You're safe at home now. There's nobody here except you and me. Nobody here would want to harm you. Everybody loves you, so just calm down and stop worrying, right?"

"Where are we? Rome?"

"No, you don't go home to Rome any more, you come here to Monrosso. Thank Goodness you remembered that. You're up by the Refuge. We've been worried sick they'd got you chained up somewhere, or you'd got stuck behind the dustbins in Peshawar again. Anyway, that's all over now. You're home, and everyone can relax. We'll have a big party tonight in your honour. Now, come on, down to the House. You must be starving. Full English today, eh?"

Cesare writhed in obvious distress. "Can't. Just need a bit of time - just happened – need time to get myself together. Hundreds of years since the last time I was blinded. Need time to adjust. You carry on. Leave me to sort myself out."

Justin stared at him, perplexed. Well, he was new to this immortality game, so maybe you didn't resurrect in perfect condition every time. He felt a strong urge to drag Cesare up onto his feet and steer him down to the Great House in triumph, but he knew he didn't have the strength to make Cesare do anything he didn't want to, even if he was blind. But would he be safe here all alone? Might he try to move and injure himself? Might the animals trample him? The wolves eat him?

The wolves answered that question for him, appearing out of nowhere and jumping all over Cesare, whimpering and licking him all over. A lot more wolf to wash off, thought Justin. But their affection for him was very evident. After all those years apart, the parents still loved him as their saviour and the three remaining cubs took the lead from their parents.

"I'm sure you'd like some tea," he said, trying to sound cheerful and normal. "Five minutes and you'll have some." He rushed into the refuge and switched on the new hob. The new solar panels seemed to be working, but, of course, no kettle, so he had to use a saucepan. He stuck a pair of scissors in his belt when he took out the tea and biscuits.

Cesare was hugging the wolves and his shoulders were heaving. Justin had never seen a brave man cry. It must be the measure of the horrors he'd just been forced to endure. He blessed his lucky stars that he'd never had to endure such things himself.

"Tea up!" he said brightly. "Now, I've got the scissors. Do you want a Don Juan again, or are you starting a new fashion for Moses fungus?"

"It's okay, I'll do it my - well, maybe a Don Juan if you like."

Justin tried to use the beard trimming as an excuse to get Cesare's arm away from his face, but it was like trying to move an iron girder.

"Why don't you try my sunglasses? They'll protect your sore eyes. Try them out and see. You'll lose the use of that arm if you keep it immobile, won't you?"

Cesare removed the arm but screwed up his eyelids so hard it was impossible to see whether he had any eyes or not, but he put on the sunglasses and relaxed a little.

"What about you? Don't want you getting eye strain."

"It's quite early morning, so it's not too bright yet. I've got another pair down at the House. Now, you look your usual heart-throb self, so shall we go down to the House? Can you walk, or do I need to go fetch the Jeep? We ought to get you to A and E as soon as poss."

"What use is A and E? They don't have spare eyes, do they? You couldn't get someone to bring me up a sandwich, could you? Den or Iska will take this in their stride - they're pretty battle-hardened, but not Lekrishta. She's suffered enough. I left her to sort out the mess all by herself, as usual. Was I in time, or did the tax people foreclose? I had no end of trouble finding the rats."

"Don't worry. Gave us all kittens, but you did it." It seemed unwise to burden him with the frightening truth:

that the bounty money may have gone astray.

"Poor Krish!"

"Well, all's well now. Are you sure you won't let me take you to A and E? Or at least down to breakfast."

"Sorry, don't think I can face people just yet."

"Okay, then, I'll go get you a sandwich. Don't go away. Oh, better get you some clothes first. Don't want you frightening the donkey, do we?"

Den and Iska just nodded, tight-lipped, then did exactly as Justin suggested. While he drove back up quickly in the Jeep, giving Cesare little time to do anything drastic, the other two walked up so that Cesare had time to eat his sandwich in peace. They crept up silently, sat down on the grass and listened.

"Has Krish been able to pay everybody yet? I suppose the money may have only just arrived."

"No, but she's managed to avert two bankruptcies. She worked out that there are anomalies in the tax bill and persuaded them to let her pay thirty per cent now and go to a tribunal about the rest. Then she talked Tekoverde, you know, the firm that supplied all our marvellous new equipment, into taking twenty per cent now and the rest in a month. She was brilliant at playing for time."

"Well, if there's anything left over it can go to the field hospitals."

"Mmm," said Justin, "suppose it can." But there won't be any left, he thought, even if the Yanks pay up. This place just isn't viable. Even back to square one, with all Albertelli's debts paid, this place would be in the mire again in no time at all. No wonder that Manager got in a mess. He was trying to do the impossible. How do I tell them that? It's not really my business, is it?

"So, are your eyes feeling any better, now you've taken on a bit of fuel?" asked Justin.

"Don't have any eyes. I told you that."

Den and Iska exchanged glances and raised their eyebrows.

"Hello, ChesRa. Welcome back. We're having a big celebration tonight to thank you for saving us," said Iska.

"Who's that?" asked Cesare.

"You know perfectly well who it is. Open your eyes, you dope." He whipped off Cesare's sunglasses.

Once again he screwed up his eyes so tightly it was impossible to know whether he had any eyeballs or not. Justin could see no blood leaking from under the eyelids before he quickly put his arm across his face.

Justin took the glasses from Iska and put them into Cesare's other hand. He put them back on with gritted teeth.

Iska opened his mouth to say something, but Den put a finger to his lips and shook his head. Iska replied with a helpless gesture.

"Hello, ChesRa," said Den. "Welcome back. You've really done us proud. Everybody has been worried sick about you, so you'll have to put them all out of their misery and come down to say hello. We've got the Jeep here and we'll take you down. No, you don't need to see where you're going. We promised not to throw each other in the lake, remember?"

Cesare looked near to tears again as they frogmarched him to the Jeep and heaved him aboard. It was certainly tough love, but perhaps the best way to treat a shell-shocked warrior. Gentleness and sympathy might be difficult for him to handle. The love the wolves had shown him had been too much for him to bear.

They marched him unceremoniously into the Morning Room and dumped him into an armchair.

"Now, Chez," said Den, "you're facing the window. The fireplace is to your right and the big table is in the middle. When you feel like trying to get around on your own, here in the Morning Room should be an easy place to start. Lekrishta has ordered everybody to attend lunch at one and dinner at seven-thirty in here every day, so we can keep each other in the loop. I don't know if you are happy to continue these arrangements. Perhaps you could talk to her. We've been having breakfast informally in the kitchen, to save time and work, as we've all been run off our feet with new things to do."

"Do you have a lot to do today?" asked Cesare. "Where's the best place for me to keep out of your way?"

"No. All our ventures have suddenly dried up. We were planning to spend the day tidying up and taking stock, so you've come home at exactly the right time. You need to decide what you want us to do next," said Iska.

"But no hurry," said Den quickly, seeing Cesare's exhausted expression.

"Some more breakfast might be a good idea," said Justin. "You've only had one sandwich. I'll go ask Cook to make you another."

He rushed off, ignoring Cesare's insistence that he was not to bother.

Cook had to fry up the filling, which took a few minutes, and by the time he returned, Cesare was alone. He was making his way slowly around the room, reaching out blindly for the walls and furniture. It was a relief to see that his eyes seemed to be his only injury.

"Bet you can't find the table," Justin joked. "You can have this egg and bacon sandwich if you can."

"You're on!" said Cesare, with a good attempt at a grin. "Within two minutes?"

238

"Sounds reasonable," he agreed.

It was a pretty good impersonation of a blind man: certainly convinced Justin. The smell of the bacon must have inspired him, as he strode determinedly around the room, stumbling against obstacles, and finally ran his hands along the table top and whooped.

"Found it. How long was that?"

"Ninety-two seconds," lied Justin, who had made no attempt to time him. Why bother? Blind men are easily fooled. What a sickening predicament for a great warrior!

"Come on, I've earned a butty," the warrior grinned, sitting down and thumping his fist on the table.

"Here you are; don't say I never give you anything. I'll get you a napkin. You're sure to get egg on your face."

The servants were their usual cheerful selves, so it was obvious that Den and Iska had not informed them. Presumably Cesare had suggested the pair of them get on with their tidying up and leave him in peace, and out of helpless embarrassment they had gladly scarpered.

It was Lucrezia's job to tell the staff, of course, but she was presumably down at the Lodge, wrestling with the missing bounty. Somebody should tell them, before they barged in on him and got a terrible shock.

The Housekeeper seemed too shocked to reply, just gazed at him, wide-eyed with horror. He spotted a pile of clean napkins and hurried out, leaving her to spread the word. When he went back later in search of tea everybody in the kitchen was crying, and he had to make the tea himself. Oh, dear! It was worse than a funeral.

"Morning everybody," chirped Kerallyn. "Am I too late for breakfast?"

"What would you like?" asked Justin. "I'll make it for you. Everybody else is hors de combat."

"What's happened? Has somebody died or something?"

239

"Cesare's come home blind. Says they gauged his eyes out."

"But that's not possible, he's -"

Justin stopped her just in time, and steered her out of the kitchen. "Come and talk to him yourself. He needs somebody to keep an eye on him till he learns to cope, and Iska and Den seem to have chickened out."

"What about Krish?"

"Haven't seen her this morning. She must be down at the Lodge trying to sort things out. And it would be cruel to tell Cesare the bounty hasn't been credited to him yet. He insists they send it beforehand so they can't do the dirty on him after he's got their man, but Krish thought he'd given up hunting rats, so she closed the account. The bank has probably sent the money back. Poor Krish. She must be feeling frantic."

"Oh, Pterodactyls! What a horrible mess! And we were all feeling so pleased with ourselves last night."

"Could you suggest some medical help for Cesare? You said you were a doctor. Did you mean just a PhD?"

"That as well. I like studying, so I keep doing courses. Medicine pays very well in the States, so a few years practising that pays for my next course. Did you say they'd gauged his eyes out? There's no treatment for that. How can there be? They can't supply new eyes, can they?"

"So there's no point in taking him to A and E then?"

"None at all, unless the sockets are infected or the bleeding won't stop. I'll go have a look."

Cesare, as if he had eyes in the back of his head, turned as she walked silently towards him, and stretched out his hand. "Krish! Krish?" He looked puzzled. "Krish? Are you alright?" he asked hesitantly.

"Hello, Cesare. Delighted to meet you at last. I've

240

heard so much about you. I'm Kerallyn. You don't remember me, I suppose, after all these thousands of years. Last time I saw you you were an absolutely gorgeous-looking black man. Now both of us are white. Weird, isn't it?

"Kerallyn? You're Krisha's twin? You certainly have a very similar aura. No wonder you had me confused!. What do you look like these days?"

"Very much like Krish, as always. My hair is shorter, and my New England accent sounds different."

"So, you've lived in the States?"

"Yes, for several lifetimes. I'm here on secondment to Sapienza University's genetic research labs. I want to have a good long look in secret at our genomes, see what on earth has gone wrong with us."

"Fascinating,"said Cesare."Do you have any theories?"

"The heat shock response could hold the key. Almost every living thing throws the switch to turn on ageing as soon as it becomes capable of creating off-spring. Some creatures die almost immediately after their first babies are born. It takes humans years to die, but once the switch is thrown, the body gets more and more careless in repairing itself, and lets more and more rubbish pile up inside it until it grinds to a halt."

"So, this switch will turn on in our - their – forties - fifties?" asked Justin.

"No, no, at puberty, at about eleven or twelve years old."

"So the poor humans are condemned to a very slow death from the age of eleven," breathed Cesare. "Poor things! No wonder they're so keen to find an immortality treatment. So, you are going to help them?"

"Do you really think I should?" she demanded. "No, I'm looking for a way to turn my switch on, to see an end to this interminable life at last."

241

"I'm very late for breakfast. Is Cook in a good mood?"

Lucrezia! Oh dear! Somebody should tell her -

Den appeared from nowhere and dragged her around the corner out of earshot. A few moments later she walked into the Morning Room looking puzzled.

With a grim face, Cesare held out a hand towards her, seeming to sense her presence. "Krish, I'm so sorry. I left you with all the problems - "

"Sweetheart, that's what you always do. We each do our jobs: you go off to fight the foe and I keep the home fires burning. We're a good team, aren't we? Thank Heavens you're home! I've missed you so much." She drew his head against her chest and pressed her face into his hair.

"Let's go make Krish some breakfast," said Kerallyn, grasping Justin's arm and steering him off to the kitchen. "Oh dear, oh dear, oh dear!"

By the time they returned with a sandwich for Lucrezia she was already hard at work on the in-house telephone. They were right not to make her a sit-down breakfast.

"Yes, that's right, shapes you can read with your hands, with your eyes shut. Letters for the main rooms, maybe, 'M' for the Morning Room, 'K' for the kitchen. We'll nail them to the door posts. We'll do the Great House first. Thanks very much, Roberto."

"And the Dairy. I can help with the milking and the harvest."

"How are you going to mark the fields? They don't have door frames," asked Justin.

"Mmm, finding that turnip field is going to be a challenge," said Cesare. "Could fall in the pond – and the muck heap."

"Well, do them in the right order," said Den. "Muck heap first, then the pond to wash the muck off."

"The ducks will laugh their heads off," laughed Cesare. "Ever heard ducks laughing, Justin? They seem to have a great sense of humour."

In the kitchen, looking for more tea, Justin exploded.

"Stiff upper lip is one thing, but he's genuinely laughing. Must be hysteria. How can a man feel cheerful when he's just lost his eyesight? He's flipped his lid."

"Isn't it obvious?" said Kerallin. "Wouldn't you be happy if you realised nobody could expect you to go to Syria again, killing those psychopathic fanatics? Didn't you tell me he'd been tortured to death several times recently, as slowly and painfully as possible? He must have passed his limit. His unconscious mind had to find a way to protect him from any more agonising torment."

"You mean he's putting this on? It's just a pantomime?"

"Not consciously. I guess he sincerely believes he can't see. His unconscious mind may have found a strategy that works. It will maintain it until it judges that the threat has gone away. The happier we make him the sooner he may get his sight back, unless he really has lost his eyeballs. Since his body is immortal that doesn't seem possible. I need to have a look, but it may be very hard to persuade him to open his eyelids."

"You sound like a psychiatrist."

"I am. It's my field. Rich pickings in the USA. Everybody thinks they're screwed up and need a doctor to sort them out. Krish is doing exactly the right thing. She's not asking questions, just accepting what he says, giving him lots of affection and helping him find something useful to do. Work is the very best way to cure psychiatric problems, to convince your unconscious mind that you are some use to the world, to give you back your self-respect. We should think up lots of ways to help him get around without help, so he doesn't feel a burden to us all."

243

CHAPTER 31

TAKING STOCK

Lucrezia had spent part of the morning writing cheques to all the 'Little People', the ones who had accused her of driving them bankrupt. Over lunch she relayed the information like a confessional, and Cesare sighed with shame.

"Our neighbours looked up to us, and we let them down. We don't deserve this place," she said. "Maybe we should sell to someone who'll make a better job of it."

"Someone who'll sack all the staff and turn their homes into holiday lets? Who'll keep thousands of animals chained up in sheds? Someone who'll shoot all the wildlife and poison every insect and wild flower? Could you really walk away and let that happen?" asked Iska

"And how will you spend your time? Go back to Syria or Iraq, trying to save the world? I've concluded that it's hopeless," said Den. "They're all so angry and bitter. Every killing makes them hate each other more. There's nothing outsiders can do, is there? I've been caught up in conflicts like this in loads of different countries over thousands of years. Surely you have, too. It's a hideous, self-destructive disease that breaks out whenever there's a glut of young men. Maybe it's Evolution's way of trimming the hot-heads out of the population and it just has to run its course."

"If you feel in need of a challenge," said Justin, "try sorting out Monrosso. These last three weeks have been a roller-coaster ride. It's a pity you missed it, Cesare, but your problems are not over, not by any means. You may think once you've paid off all the debts you'll be living in clover, but I give you six months, maybe even less, and your debts will be mounting again. This place is barely breaking even. The slightest problem will send you into the red. If Albertelli was an honest man he'd have told you he couldn't balance the books. Maybe he cleaned you out and ran to avoid being blamed for doing a bad job and getting fired. Am I right, Lucrezia?"

"Yes. Now I've got the genuine bank statements I can see we were in trouble," she sighed, "even without the trips to the betting shop."

"How on earth can that be? We've been doing fine for hundreds of years. We've never gone bust before," Cesare protested.

"Things have changed. Farming doesn't pay the way it used to. We've already cut out most of our energy costs, gone green and self-sufficient. We've done all we can to the vineyard. It's as modern and productive as we can make it, but the market's drowning in wine. Even the British are growing their own nowadays. You need to diversify," said Iska.

"You'd need another big cash injection," said Justin, "to set up lots more money-making ventures."

Cesare groaned.

"That's not a dig at you, Chez. You've done more than enough horrible heroics to last hundreds of years. You should do things the normal way in future. Nice and businesslike," said Den. "We've made a start, hosting special occasions, but we don't have the right facilities: no cloakrooms, no flower garden, no furniture and stuff."

"We've tried to open a farm shop, but we need to spend money to get it past the inspectors," said Iska. "And we want to spin our own alpaca. Make far more money, but we need the machinery. We're selling direct to lots more delis now. Much more profitable than selling to the farmers' co-op. But we need a refrigerated van to carry it as far as Milan. Then we'd really make money."

"I don't think we're up to it. We'd better admit defeat

and let someone more able take it on," sighed Lucrezia.

Justin looked at her weary, drooping shoulders, and at Cesare, still wearing his sunglasses.

Poor Lucrezia! Cesare was just one burden too many. His arrival had helped her revive the Cayman Islands account in both their names, but she was now spending hours chasing the bounty that the bank had returned to sender. There was no guarantee that it would be paid a second time. She had sworn everyone to secrecy about their continued plight. Cesare had suffered enough, she said. Imagine having to take on board the realisation that all his suffering had been in vain, that the bounty had not been paid! She had three more weeks before her agreement with Tekoverde expired. She hadn't given up trying, but she must be preparing him for the worst.

After lunch Lucrezia led Cesare off on a tour of the kichen garden, to tell him about their new cafe and farm shop. Justin cornered the others and proposed a council of war.

"Before we go any further," he said, "Would you be good enough to tell me honestly where I stand, now I seem to be an Immortal, like you?"

"Yes, I want to ask you the same thing," said Kerallyn. "You've been very kind and welcoming, but -"

"Well, of course we have: you're family," said Den. "Families as weird as ours should stick together. We need all the allies we can get."

"Am I an ally?" asked Justin.

"Well, I sincerely hope so," said Iska. "You're too clever by half. I'd hate to get on the wrong side of you."

He's just being sarky, thought Justin, but there was no mockery in Iska's expression. Amazing!

"Well, then," he said, "If I'm one of the gang maybe you'll tell me what you're planning. Cesare's blindness looks like the last straw for Lucrezia. He won't be much help running this place, will he? Are you thinking of advising her to sell up and find a nice little bungalow by the sea - "

"What?" They all shouted at once. "Lose Monrosso! Over my dead body!" - or words to that effect.

"So you're hoping to stay on, then, now Cesare's back home for good?"

"I certainly hadn't thought about leaving," said Den. "But I don't know if they're thinking of turning me out."

"Turning us out?" said Iska. "I never thought of that. But this is home. Where would I go?"

"So," said Justin, "I gather you want to stay."

Both frowned and nodded.

"Well, as both Cesare and Lucrezia seem to be at a very low ebb you could act now. Ask to become share-holders in this place with a right to live here. I'd welcome a chance to do that myself."

"So would I," said Kerallyn. "This place is a Garden of Eden. It's been so much fun playing Save Monrosso that I can't bear to think that I might never get invited here again. And I think we make such a good team."

"Couldn't put it better myself," said Justin. "And I've an idea I'd like your advice on. I need a new headquarters, part design team, part emergency flying squad. The Lodge would be perfect. If you don't intend employing another estate manager you won't need it, so I could buy it. Then we'd have a couple of million to spend on setting up lots of new ventures, wouldn't we? I've been wondering if the garage of the Lodge would make a good farm shop. It's close to the entrance, so we wouldn't have people driving around the grounds, knocking the animals over. When there weren't many customers around, my staff could open up the shop on demand."

"Well, if they were selling the Lodge to a stranger I'd be pretty unhappy, but you're family, so that's different."

Whihee! thought Justin. Iska thinks I'm family. Maybe I can stop watching my back.

Reluctant to tear themselves away from Monrosso, they decided to leave the journey back to Rome till Monday morning. They could enjoy a relaxing Sunday dinner with the gang and celebrate Cesare's return.

After dinner they all sank into comfortable chairs in the sitting room and Lucrezia put on 'Capriccio Espanol', both cheerful and relaxing.

"Now, let's have a tasting test," she said. "This is our new Marquis Special Reserve brandy. It's selling very well and getting good reviews. All this expensive new equipment Cesare has paid for seems to have improved it no end. Tell me what you think of it."

They sipped it dutifully, and murmured their approval. Very pleasant, sitting in the dusk, listening to romantic music, watching the sky grow darker and the evening star grow brighter by the minute.

"Cesare!" called Kerallyn. "There's something wrong with Cesare."

Lucrezia had insisted he should sit on the sofa, but he appeared to be in danger of rolling off it.

"No, don't put the light on," Lucrezia insisted. "Yes, give me a hand, Iska. Get his legs up on the sofa. That's right. Pull him down so we can get his head flat'"

Justin watch in some alarm. What on earth was going on?

Lucrezia took off his sunglasses and gently opened his eyelid. "Den, shine me a light in his eye."

Den held his hand over the eye, and as the hand began to glow, he held it fingers down, concentrating the beam into Cesare's eye.

"Take your hand away for a moment. Come and watch

this, Iska. What do you think?" She nodded to Den to bring his shining hand back into play. Then she repeated the procedure with the other eye. "Verdict?" she asked.

"Psychosomatic," said Den.

Iska nodded. "Shell shock."

"May I look?" asked Kerallin.

"You a doctor?" asked Den.

"Ain't evrybardy?" drawled Kerallin. "Yipp, reckon you've got the right diagnosis, thank Heavens! A clear case of Conversion Disorder, formerly known as Hysterical Blindness. But it could take years to let up, and in that time his brain might lose the ability to see."

"Troubles are like buses, always come in fleets," sighed Lucrezia, "but thank goodness this makes sense."

"We could kill him and see if he wakes up okay," said Iska.

And Justin caught himself thinking that might be a good idea!

CHAPTER 32

HEN PARTY

"Thanks for all your help last week, Freddo. You got me out of a pretty embarassing situation."

"So, you had to play, then?"

"Yes, I played an hour and a half of Gershwin in front of about two hundred people. Had the time of my life. Where did you get your drum kit? Going to get one myself. Now, how's that snagging going? All done now? Working better? Fine. Now, listen everybody, how would you like to work up beside Lake Como? Or are you all too attached to the city lights?"

"A bit far to commute, isn't it?"

"Well, obviously you'd have to relocate. Gather round and I'll show you some pictures. Just kick it around for a bit, let me know how you see it."

His penthouse seemed strangely quiet at first with only one gorgeous redhead in residence. Lucky they had all been at Monrosso the weekend Marcella's husband was away! Three gorgeous women would have been two too many. The time had come to wean himself off Marcella.

Kerallyn made no attempt to move into Lucrezia's empty room, nor did she ask about his vacant tourist flats. She had settled into his bed like a little dormouse, all cosiness and charm. It would be cruel to throw her out, wouldn't it?

Lucrezia had rebuffed all his advances, so there was no reason at all, surely, why he shouldn't transfer his attention to Kerall. She didn't seem to have any reservations, did she? And if she thought he was boring in bed she never even hinted at it. She certainly wasn't!

Her thousands of husbands had taught her the most amazing things.

Poor Lucrezia! What was her reward for her devotion to Cesare? Constant hard work and anxiety. Identical twins but such different personalities!

"What are you up to, Kerall?"

She had suddenly begun to do some rather odd things. She wore a constant frown, and was spending her evenings in his spare room, working on her computer, and occasionally rooting about in the wardrobe. She had been similarly preoccupied all weekend.

"I'm off on a trip on Tuesday," she said. "Back about the middle of next week."

"Oh, really? His heart sank at the thought of a weekend all alone. She was such a cosy little fixture. It would be very hard to watch her go off back to Boston when her secondment came to an end. "Are you going far? Can I give you a lift?"

"No. Thanks for the offer, but it's all arranged."

"So, where are you going?"

"No idea. It's a girls' weekend. Top secret. One of the girls at the lab is getting married in a few weeks."

"Ah, it's a hen party. I see. Do you all have to dress up in silly costumes? I see you've got a couple of scruffy-looking arab costumes in the spare room wardrobe." Well, it is my spare room, he thought. And it's not as if she's sleeping there. Some of Lucrezia's frocks were still hanging there as well.

"Mmm. We're supposed to look like arab slave girls."

"Slave girls? That's a bit off, don't you think, with those terrorists in the Middle East making sex slaves out of all the Yasidi women they capture? Who's idea was that?"

"Look, it's not my wedding, so I just have to go along

with it, don't I? Let's drop the subject, shall we?"

She looked so cross he shrugged and went off to make some tea, determined not to raise the subject again. But on Tuesday, at the last minute, he decided to take the afternoon off to wish her a good trip.

As his penthouse doors swung open he recoiled in shock. Two figures were bustling about, two repulsive-looking, scruffy old arab ladies. What on earth were they doing in his flat? Had the hen party invited their grannies as well?

"Kerallyn!" he shouted. "Kerallyn? Where are you?"

"Oh, you're home early," said one of the figures in a Boston accent.

"Hello, Justin," said the other identical granny. "Hope you don't mind my touching down here for a moment."

Justin stared at them in disbelief. Both had dull dark hair touched with grey and swarthy skin covered in wrinkles and warts. Their smiles revealed ugly stained teeth, and they both smelled stomach-churningly musty. A memory of Cesare in a masculine version of their costume leapt into his mind. A genius for disguise must run in the family.

"Kerall? Lucrezia? You look absolutely amazing and disgusting."

"Every man's fantasy slave-girl, eh?" laughed Kerallyn.

"You - ! You wicked little - !"   He burst out laughing.

"Anyone who bought us all covered up in niqabs would get a nasty shock, wouldn't he?" laughed Lucrezia.

"Serve him right. Do we look authentic?" asked Kerall.

"I've never met an arab slave girl, but you look disgusting enough to win first prize in a fancy dress show. Well done!   So, where are you off to?"

"Big secret. See you some time next week. Now, have we got everything? Could you just sit down quietly for a

252

moment, Justin, so we can concentrate."

"Of course. Ooh, those look good." There were lots of little packets on the table, full of enticing lumps of fudge in delicate pastel colours, and chocolate wedges, all with dainty decorations."Your home-made sweets? Are these arabic letters?"

"Yes. They mean 'best of luck'."

"You can spare a little packet, can't you?"

"No! No! Don't you dare touch those!" they shrieked in unison. "There's just one each for everybody. Can't you see the labels? We can't leave anybody without. Put that down. PUT IT DOWN!"

"Crikey Moses, keep your wigs on," Justin protested. "All that fuss for a few sweets." He sat down sulkily and watched them pack the sweets into their string bags.

"Right, have we got everything?" asked Lucrezia. "Got your niqab? Wish us luck, Justin."

"Let me give you a hug. Good luck. Hope you win a prize. Where's your luggage? I'll carry it down for you."

"This is all we need," said Kerallyn, swinging the tatty string bag. "See you next week. Look after yourself."

Both 'old ladies' gave him a hug, then strode off, leaving his flat smelling like an old crone's drawers.

Justin made a big mug of tea and rooted out a few Ginger Nuts, to deaden the dismal feeling of being ditched by two gorgeous red-heads at once. He reran recent events in his mind and tried to make five out of two plus two. What were they up to? Had Lucrezia really been invited to the hen party of someone she was unlikely to have met? And why had Kerall recently spent so much time on her computer in the evenings? She had all day to pursue her research, and for that she needed the gene sequencing machines in the lab.

Suddenly it occurred to him that she had left her laptop

behind. He set it up on the dining table. "I'll soon find out what you're up to, you wicked little witches," he chuckled. "No computer can hide anything from me!"

It was all too easy to spot where they were going. The deleted emails were in the computer's recycle bin. There were not only flight bookings, there were boarding passes too. Two single tickets to Baghdad, via Istanbul. Single tickets! Good Grief! How were they intending - *were* they intending to come home? A cold shiver ran down his spine. She had studied Fallujah on Google earth. It was only a few miles from Baghdad airport. Fallujah, Islamic State, so terrifyingly close to Baghdad they shared an airport.

Newspapers. She had keyed in so many newspapers, Iraqi, Syrian, Kurdish, Turkish, Saudi, Jordanian. What was she looking for? It was Lucrezia, all over again, trying to keep tabs on Cesare as he chased his murderous rats. But who was Kerall trying to keep tabs on? Surely not Cesare. He was hors de combat. Den or Iska? They may not be prepared to try to save the world, as Cesare had been doing, but they were very keen to save Monrosso, and battle-hardened, according to Cesare. Had they gone to Fallujah, and if so, who was left to run Monrosso? One blind man? Were they locked up in Fallujah? Had the twins gone to rescue them?

He needed a run to clear his head, after all those hours at the computer. After once around the block the penny dropped. Idiot, he almost yelled aloud. Would Lucrezia try to catch a rat? Of course she would! Anything Cesare could do she might try to emulate. Kerall had led a much less challenging life, with husbands who weren't heroes, but she was a twin, shared Lucrezia's DNA. Maybe she would relish the chance to show her twin she was a heroine too.

But they couldn't kill with a Karate chop to the throat, not with their dainty little hands. They'd need to find a

254

weapon, but how would they get that past the guards? Cesare had said the king-pins who directed operations were often cowards who surrounded themselves with bodyguards. Cesare was used to the rough-stuff, had been a warrior for thousands of years. They were fragile little creatures. Heaven knows what would happen to them. Torturing them would be like pulling the wings off butterflies. Oh my God!

It was a horrible week. There was nothing he could do. Should he try to get in touch with the family at Monrosso? Maybe they had kept Cesare in the dark. It would be cruel to alert him, present him with an agonising dilemma. He'd be certain to blame himself. They must have bamboozled him with their tale of a hen party. Better leave him believing that for now. Den and Iska? What could they do? They'd feel obliged to do something, which could make matters worse.

Maybe it was insulting to worry about them, suggesting he considered them unequal to whatever it was they were set on doing. With twenty four thousand years of experience between them, surely they must know what they were doing.

The news broke on Saturday. According to the world's media, a dreadful tragedy had struck a party of revellers celebrating the end of Ramadan in Fallujah. It was the Friday of Eid, a great three day celebration of feasting and present giving. The party were enjoying the traditional breakfast of sweetmeats when they were suddenly taken ill, so suddenly it could not have been an infection that killed all seven people – or five, according to later reports.

No wonder they threw a fit when he tried to eat those sweets! Lucrezia Borgia! Didn't she have a reputation for poisoning her enemies? He'd chosen to believe that was only propaganda, but now -

He was shaking, and the hair was standing up on the back of his neck. Calm down, Idiot. It's just a coincidence. Why blame the Redheads? It could be anybody; and the victims could all have been incubating something for days. And surely it was over-kill. Had all seven of them had prices on their heads? Would the pair kill indiscriminately, both the monsters and the harmless?

For the rest of the day he trawled the Middle East newspapers, devouring everything he could find in English. Soon names were quoted, then details. He began a list, and tacked on everything he could discover about every one of the victims. All were men, and all seemed to have a large number of scalps to their names. Was ever such a collection of villains executed so expeditiously? Well done, Redheads! Thank Heavens they'd stopped him eating those sweets!

But now what? Had they been caught? Were they chained up in prison, being tortured. Surely now he should alert them at Monrosso, so they could try to organise a rescue.

The phone rang. Probably some crook in India trying to tell him his computer had developed a fault and they could fix it for a fee. They always rang about this time, and they'd keep ringing till he answered. They got his standard answer. "I don't have a computer," he growled. Clear off!"

"You've got loads and loads of computers, Fibber," said a woman with a New England accent. "What are you up to?"

"What am I up to? You little - ! Where the hell are you? I've been worried sick."

"Having a girls' weekend. We told you. We got bored so we came back to Monrosso. Just ringing to be friendly. Are you having a good weekend?"

"Brilliant," snarled Justin. "I've got the flat full of gorgeous blonds."

"Oh well then, I won't spoil your fun. Chow."

"Hey, don't ring off!" Silence. Idiot! Why did you say that? Ring her back straight away. Of course he didn't have her number. Monrosso still had no land line and her mobile was on his bedside table. Idiot man!

CHAPTER 33

DEADLIER THAN THE MALE

When he arrived home on Monday evening she was making the supper. Actions seemed better than words. He simply grabbed her and hugged her till she squealed he was breaking her ribs.

"What did they do to you? I've been worried sick about you. Thought they'd be torturing you to death."

"Well, you needn't have worried. We just waited till we were sure they were as good as dead, then we ate the rest of the evidence. Cyanide is not too bad a way to go, if the alternative is struggling through three big airports and suffering two flights on Middle East Airlines. And those sweets were really delicious."

"You look as fit as a fiddle, thank Goodness! You looked disgusting when I last saw you. It must have taken ages to scrub off all that dye."

"No need for any scrubbing. The dye stayed behind in Fallujah when we died, along with the clothes."

"You don't seem the least bit traumatised. You've ended the lives of seven men. Doesn't that mean anything to you at all?"

"Seven? They must have counted us as well. Justin, they were monsters. They'd be condemned many times over by any war crimes tribunal. If only you'd heard them congratulating each other on the most horrible atrocities, carried out under their orders: endless decent, harmless people being thrown off rooftops, stoned to death, blown to bits while they were doing the family shopping. Heads cut off on public street corners. One huge fat ugly creep was gloating about how many Yasidi women he'd raped. Everybody was patting him on the back."

258

"They said all that in front of you?"

"Justin, women hardly count as human beings to them. And old ladies are invisible the whole world over. Anyway, we paid the penalty for murder, didn't we? Executed ourselves with the same poison we used on them. Howzat?" She put a dish of New England chowder in front of him. "I dare you to eat it."

How could he refuse, and retain any self-respect? How quickly did cyanide take effect, he wondered. Did it hurt a lot? It was hard to force it down at first, with her laughing at his efforts to swallow it. But it was seriously good, and seemed to be doing him no harm.

"How on earth did you manage to find so many monsters so quickly? Five at one go sounds incredible."

"Until a few weeks ago Krish worked near Fallujah with Medecins sans Frontieres. Long ago she'd lived in the area for generations but she never let on how well she spoke the local dialect. The injured fanatics never dreamed an Italian woman doctor could understand every word they said, so they talked quite openly while they were having medical treatment. She'd heard that a group of leaders were scheduled to hold a summit there over the Eid Breakfast, and then address a huge public meeting. We just had to cross our fingers that they hadn't changed their plans."

"And obviously they hadn't changed their plans."

"Not till we changed them for them."

"It beggars belief. You look so sweet and innocent, both of you. So, how did you get near them, and how did you get away with it?"

"Well, a few hours before, we walked in with brooms and pretended to sweep the place out, to make sure it was the right venue. Then, as they began arriving, we floated in with the sweets on fancy trays. We put a few at each place setting, to make sure nobody got left out.

Then we pottered about pretending to be servants till they started clutching their throats and falling about. Then everybody was rushing around panicking. They had no reason to suspect two old servant ladies - I told you old ladies are invisible. Then we just crept around scoffing all the sweets that were left, as any old lady would, then we clung together so we both woke up at Monrosso, up by the Refuge where you found Cesare. Lovely place to wake up."

"Whew, you beggar belief!" Justin whistled. "A very slick operation. Have you done that sort of thing before?"

"Of course!" she said airily, tucking into her chowder. "But I've also been murdered in every conceivable way as well, over the years, so you can't expect me to get too squeamish about a chance to get my own back a little, can you? Men freak out at the thought of women killing them, but their record of extreme cruelty to women is thousands of times worse. They surely deserved it for those poor Yasidis alone, didn't they? Raped and killed hundreds of them, just for being Christians."

"Suppose you're right," Justin sighed. "But why did you do it? I presume there must be bounties on some of those monsters you poisoned."

"The Powers That Be sent Cesare a message with three new names on, and said they would have to put his last rats back on the list as well – it was the only way they could pay him for those. The original fee the bank sent back was now sunk without trace in some sort of slush fund. He would have to see off at least one more rat to be paid for the ones he'd already done. Well, he obviously couldn't do that, could he? Krish decided to keep the message to herself and see what she could pull off in his place. She wanted somebody to know what had happened to her if things went wrong, so she told me. Of course I wasn't going to let her go off alone. I guessed you'd sneak into my computer to find out where we'd gone, and tell the others if we didn't come back."

"You took the most frightening risks. It twists my guts to think of what they might have done to you both."

"Justin, you've only been around for twenty-nine years. We've all suffered the very worst that can happen to a human countless times over. You can't expect us to see things in the same light as you do, can you? You've a lot of learning to do."

A cold shiver flickered down his spine. How did she manage to be so upbeat and cheerful?

"So," he asked, "have you made enough to pay the debts? Today was the final deadline, before Tekoverde sue us, wasn't it?"

"Krish had everything ready to pay them this morning, by credit transfer direct from the Cayman Islands. That means all our debts are cleared, apart from the rest of the tax bill, and there's oodles of money now to pay for that. We contacted Cesare's paymasters again on Saturday and asked for any other bounties on offer for the others we poisoned and they credited another two this morning, just before I left. That's twenty three million dollars altogether, including the six for Chez. There were some very nasty pieces of work among the sweet-eaters, worth five million apiece, dead. We can set up any scheme we can possibly think of now, can't we? Krish, of course, wants to give most of it to the field hospitals and refugees, so I'm keeping my half separate, ready to step in if they bankrupt themselves again by doing too many good works."

"You two must have made as much as Cesare, and all at one go. You're sensational."

"Yes, we're sensational, sensational, that's all," she crooned. "Krish says Cesare made about eighteen million for the field hospitals and refugee camps, plus this latest six for Monrosso, so he's still well ahead of us, thank goodness! And poison isn't a brave man's weapon, is it?"

"The female of the species is deadlier than the male. Who said that?"

"Every biologist, I expect. Nature is full of examples."

"What does Cesare say about all this?"

"We haven't told him, or the brothers. We've only said his payment has come in a little late, and it's enough to pay all the bills. He's saved Monrosso. Krish is going to keep her bounty in a separate account as well, so she can drip-feed money quietly into Monrosso and the field hospitals when it's needed and play the market on the side. Cesare and his brothers are apparently not much interested in finance, so they probably wont notice. They're happy to leave the boring work to her. She's asked me to get you to promise not to tell anyone what we've done, certainly not Chez."

"Of course. I wouldn't dream of it. But I'm surprised the Yanks agreed to deal with you two. I thought they'd vet their agents very carefully."

"Well, Krish dealt with them by text and email, of course, using all Cesare's codes and passwords. They've no idea they were dealing with a woman. Anyway, we hope that's it. No more assassinating monsters. From now on we're all going to live a normal life – at least for a while. We're having a big celebration next weekend for all the staff and tenants and those people we owed money to. Hope you can make it. We don't want to do it without you."

CHAPTER 34

WHAT A SWELL PARTY IT WAS

"Such a pity about the weather," said Lucrezia. "We were planning a barbecue."

It was a grim September evening. Rain streamed down the windows, increasing the joy of being safe and dry indoors. It was always summer in the ballroom. A dozen chandeliers twinkled beneath the heavenly blue ceiling, where angelic painted creatures floated amongst fair weather clouds.

"Oh no, My Lady! It's a change to be indoors, and I've never seen the inside of the House before. It's amazing," said the stockman.

"Oh what a shame!" said Lucrezia. "Come up here tomorrow, bring the family, and one of us will take you on a tour. There's lots of lovely art to see. I love your costume. Everybody's found such fun things to wear. You've all cheered up a dismal evening. Gio's waving to you, look."

"We're going to sing. 'Sing for your supper,' your lovely invitation said, so that's just what we're going to do. We're on first. Wish us luck, My Lady."

"I can't wait. You'll be brilliant," Lucrezia grinned.

And they were – after a couple of false starts. Pretty well all the male denizens of Monrosso had joined the choir, even the Triplets, hiding on the back row, singing quietly so they didn't overwhelm the mortals.

"Where did they get those hilarious songs?" Justin asked Lucrezia. "A bit risqué for a family audience."

"These kids aren't namby-pamby city kids. They're growing up on a farm. They don't need no edicashun. They can watch the animals doing what comes naturally.

These songs go back hundreds of years, back to when nearly everybody lived on the land, close to the animals. Kids learn the songs from their elders. The same with these dances. Come on, Justin. Stretch your legs."

She dragged him onto the dance floor. "Don't worry. The others will sort you out, just see."

And sort him out they did. Justin bumbled around the floor, totally bewildered, as the dancers, doing some very complicated foot work, wove intricate patterns up and down the room. The women shrieked with laughter, reached out and grabbed him and pushed him off in directions he least expected. They were obviously having great fun at his expense. Grin and bear it seemed the only acceptable response.

Den and Iska were as conspicuous as he was. By popular request they were, like him, resplendent in their ancient wasp livery. Cesare moved around the floor quite confidently, with a housemaid, who must have raided Lucrezia's attic wardrobes for her gorgeous dress, on each arm. Lucrezia had got her wish: he looked like Louis The Fifteenth, wearing that fabulous Eighteenth Century outfit of cream brocade, covered with gold roses, and the frilly shirt. Pity about the sunglasses.

Then it was the ladies' turn to warble. They'd chosen popular songs from the Hit Parade, just to show they knew what the world outside Monrosso was up to. Kerallyn and Lucrezia joined in, causing an excited murmur in the ranks. It must be the first time many of them had seen the Marchesa's twin. Twins as well as triplets. Were all ancient Mrushans so identikit, Justin wondered. Was that significant? Were they clones from a factory on some other planet? Well, their personalities were not off a production line, were they? Each was unmistakeably unique. Food for thought.

And talking of food: people he had never seen before were bringing in salvers covered with exotic canapes and every kind of tempting food one could wish for.

"Krish has called in first-rate caterers so our staff can be waited on for a change," said Den. "Look at Maria Lepanto. She looks like Marie Antoinette, and so do Krish and Kerall."

After supper Lucrezia led Cesare to the centre of the dais and he held up his hand for silence.

"You sound to be having a fun party," he began, "either that or a riot. I hope it's a party as we need your goodwill from now on, even more than we did in the past. You all know Monrosso has been through troubled times, and we owe it to you, and to my wonderful family, that we can now look ahead with confidence.

"I hope you all have an appetite for change. To ensure a sound future for us all we have to move with the times, set up new ways to generate the income we all rely on. We've made a start on developing the Great House as a venue for special occasions and we plan to open the House and grounds to the public. The admission charge and the new cafeteria should help pay for essential repairs. The village green could become a miniature fairground for visitors' children. All residents would have free use of the amenities, of course. We may create a small petting zoo for visitors' children nearby. What animals would your children like to cuddle? Please let us know what you think of these ideas.

"Georgio Calvi has told me he's retiring at the end of the month. We'll miss you, Georgio. We wish you a happy future at your daughter's. I gather your future garden is nice and flat. Come back and see us in a few months' time and you'll see humans competing with the goats, trying to climb those impossible cliffs of yours. My money's on the goats.

"Yes, we plan to set up adventure training courses, using the rugged areas and the belts of trees. There'll be zip wires and wig-wam villages for bushcraft. There'll be new jobs available, and, of course, you'll be welcome to try a change of career, or do a few hours of something different at busy times. We want it all to be a big adventure. I want to reassure you all that the farm and the vineyard will remain our main priorities, so your present jobs are safe, and we most certainly do not intend to turn your homes into holiday lets.

"I'm sure you're all brimming over with ideas, so please give us your suggestions. Now, thank you all again for helping to save Monrosso. Enjoy the rest of the party."

"Three cheers for the Marquis," shouted Old Roberto, and the noise made the chandeliers tinkle.

Two Wasps stepped up to join Cesare on the stage. With Lucrezia on the keyboard and Kerall on the clarinet, the Triplets sang 'The Flea,' an amusing song for a deep bass voice, or in this case, three bass voices. Then, with a truly shocking change of voice and mood, they minced about the dais singing 'Three Little Maids from School' in high falsettos that would pass for pretty good coloratura sopranos - unbelievably inappropriate sounds emanating from three big alpha males. The audience erupted in hysterics as his brothers steered Cesare off the stage.

It was a while before Justin realised that Louis The Fifteenth had disappeared. The King, though splendidly dressed, was far less conspicuous than the Wasps. Morning Room? No. Sitting Room? No. Loos? No. Library? Eureka! Cesare was almost hidden by the high wings of the chair. He was wearing earphones. Justin slunk silently up behind him to listen. Not a sound.

"Hello, Justin," said Cesare flatly. "I'm not shirking. Nobody's missing me, now I've done my bit."

"I've missed you. I've searched the place for you," said Justin. "Are you okay? The party's going a bomb. Everybody's pleased with what you said and you creased them up with your little girl act. You should be on TV. You'd win any talent contest."

"Hah!" growled Cesare. "Thank you. Clapped out idiot on daytime TV, that's me."

"You know you really ought to get some advice about those eyes. Have you talked to Kerall? She's very switched on, very understanding. Honestly, just talking about the problem could sort things out. All your family are doctors, like you. They all think you still have eyeballs. You just need to open your eyes."

"Justin, go back to the party," snarled Cesare. He put the earphones back onto his head and put his hands over the ear muffs.

Justin stared at him helplessly for a few moments, then admitted defeat and slunk back to the ballroom. What do I know about other people's motivation, he thought. Don't even understand myself. Get back to the party. Just get on with life.

CHAPTER 35

DISASTER

"The Marchesa, the Marchesa! Find the Marchesa!" Judy erupted into the Great House like a force twelve tornado. Their peaceful Sunday afternoon recovering from the party had come to a nasty end.

"What a racket!" exploded Mrs Lepanto. "Stop the noise immediately. This is no way for a servant to behave."

Judy flung herself at the Housekeeper, screaming even louder. "The Marquis, he's hurt. It's bad, very bad. I have to find the Marchesa. She'll know what to do. Oh my God, there's blood everywhere! Call an ambulance. Quick. Somebody call an ambulance. He's going to die, he's going to die!"

Den strode into the kitchen, closely tailed by Justin.

"Where is he, Judy? Calm down now, just tell me where he is."

"Down by the gate. His leg is cut right through"

"Gate, which gate?"

"Just past the turnip field."

"The beetroot field?"

"Yes, yes! There was this big metal ploughing thing right by the gate. He couldn't see the thing. He fell right into it. I never realised those things were so sharp. Please, Mr Fermi, call an ambulance. He's going to bleed to death."

"Ring Mr Romanov, Maria, and tell the Marchesa. I'm going straight down there. You stay here, Judy, and tell the Marchesa all you know."

He was wasting his breath. Judy turned tail and ran, almost falling down the steps in full flight.

"Ambulance, Mr Fermi?" asked the Housekeeper.

"Definitely not, Mrs Lepanto. If the Marchesa wants an ambulance she will call for it herself. Tell the Marchesa what Judy said and do whatever she instructs you. The First Aid box might be called for."

"A great big kick in the pants is what the blighter needs," Den grumbled as he set off at a jog towards the fields, closely followed by Justin. "He must keep up this ridiculous charade even when he hasn't got an audience. What does he hope to gain from it? It's sickening to see him blundering about, banging into things. Oh, there's Iska."

"That idiotic housemaid of yours just went screaming past as if she's got the hounds of Hell on her tail," Iska. shouted. "Says Cesare's bleeding to death."

"She says he's fallen over your bloody ploughshare and cut his silly leg off. Why the Hell haven't you told your men to keep the roads and gates clear. Surely they know he can't see where he's going."

"Of course I've told them, but things happen, you know. That girl was screaming for an ambulance. You haven't -"

"Of course not. Don't let your people call one, either."

"How daft do you think I am? Look, he's lying down. That doesn't look good."

He looked even worse when they reached him. Judy was cradling his head against her breast and moaning hysterically.

"Judy, behave yourself," barked Den. "You know the Marquis hates a fuss. He wont like you any better for it."

Iska knelt down beside him "Now what have you done to yourself, you great big dope?"

"You tell me," hissed Cesare, through clenched teeth. Something tried to chew my leg off."

Den knelt down and surveyed the scene. "Better get

269

his pants off and have a look at the damage."

Justin peered over Den's shoulder. There was a large gash in Cesare's jeans and blood was pulsing out of it.

"Don't be daft," growled Iska. "We can't treat him here in the middle of a field, Have to get him up to the House. You, girl, we'll have that pinny off. Make a tourniquet to cut it off. Your pinny, girl, quickly."

"Don't cut his leg off, please don't cut his leg off."

"Cut the bleeding off, you daft girl, not his leg off. Get that pinny off this minute."

Judy, sobbing bitterly, struggled with a tangled knot behind her.

"Turn around," Den ordered. He yanked the string right off and passed the pinafore to Iska.

"Den, give me a hand with this. Justin, ring the House and ask Krish to get an operating theatre ready." With the slickness of a pro he wound the pinny tightly around Cesare's thigh. To Justin's relief, the blood no longer pulsed, but his stomach turned over at the sight of the sticky red puddle soaking into the ground.

"Mr Romanov, shall we call an ambulance?" Luigi's face was white and his eyes were as frightened as a beaten puppy's.

"No, no, don't let anyone do that. We've plenty of doctors up at the House. What we need is a stretcher to carry him up there. Can you rustle something up, anything that might do?

The small crowd of onlookers that had gathered dispersed rapidly in all directions, calling, "Stretcher, need a stretcher, quickly."

"Please, Mr Fermi, call a doctor," sobbed Judy.

"Listen, Judy, The Marchesa is a doctor, Mrs Wilson is a doctor, Mr Romanov is a doctor and even I am a doctor. The only one who isn't a doctor is Mr Chase, and he's a first rate computer expert."

Judy looked at him in frank disbelief.

"Listen, girl, we have all of us worked in field hospitals in war zones. We've had to treat every kind of nasty wounds you can imagine, and lots you'd sleep much better if you didn't try to imagine. No other doctors could be more experienced and better qualified to treat him than we are. If we sent him to hospital the doctors there would be a lot less able to save him than we are."

Farmers and wives came running with tarpaulins, hammocks, you name it, and, at last, a sunbed.

"Great!" said Iska.

"Just what the doctors ordered," said Den pointedly to Judy. "Now, we need another pall bearer. Luigi?"

"Don't say that!" wailed Judy, digging her fingers and her nose into Cesare's hair.

"Get her off me," mouthed Cesare, giving her head a friendly pat.

"Judy," said Den, "we need you to open all the gates. Get you hands off the Marquis so we can get him on the stretcher. Go on ahead to the gates. Don't hold us up. He needs treating urgently."

With the utmost reluctance she untangled her fingers from Cesare's hair, then, on a sudden impulse, kissed his forehead, then tore herself away, sobbing.

"Poor little kid," murmured Cesare.

"Have you been - ?"

"No, I have not," he whispered. "Are you accusing me of baby-snatching?"

"I'll believe you where a thousand wouldn't," mocked Den.

"All together now, lift him carefully," said Iska. "Get his other leg on. Yes, I know this sunbed's a bit small but it's the best we've got. Now, all together, lift."

The pallbearers winced as the metal frame of the sunbed dug into their shoulders. Estate workers rushed forward with bits of anything they thought might serve as padding.

"No, don't jog. Don't torture the poor soul. Walk as smoothly as you can," said Iska.

The route to the House seemed interminable. Cesare sported no excess fat but his powerful muscles and inhumanly strong bones made him a formidable weight to carry. They were relieved to find Lucrezia waiting outside the basement door.

"In here, please," she said. "Better down here. More private. Put the stretcher on this table. Right, now, lets see what you've done. Let you out of my sight for two minutes and look what you do to yourself."

She picked up a pair of large kitchen scissors and speedily hacked open the leg of his jeans. The others crowded round to view the damage, Their looks were grim. All gave a sigh.

"Give me the coup de grace," Cesare murmured quietly. "Quickly, please."

"Too public at the moment," murmured Iska.

"Mrs Lepanto," said Den, "would you kindly take everyone upstairs. I want you to put up a notice that only the family are allowed down here below stairs until further notice. I think Judy and Luigi deserve some tea and cake. They've both been very helpful Thank you both. Luigi, take Judy upstairs. That is an order."

Judy stood rooted to the spot. Iska glowered at Luigi, so the boy took Judy firmly by the arm and propelled her out of the half-built cafeteria. They could hear her sobbing as she stumbled up the stairs.

Lucrezia beckoned the Immortals into the pantry and closed the door. "Now," she said, "I don't need to tell you that this looks pretty nasty. Most of the flesh has gone. We'll have to amputate. Can we lay our hands on some anaesthetic?"

"What's the point of operating? He's asked for the coup de grace," said Iska. "Good thing too. His eyes should be back to normal as well when he wakes up tomorrow."

"You really think so?" said Lucrezia.

"Of course, why not?" they choroused.

"I wish I could believe that. What if he wakes up lame as well as blind? Yes, I know that's never happened before, but all good things must come to an end. He's been in such an odd mood these last few weeks. He's shut me out of his mind and he just wont talk about it. Haven't you noticed he's not himself?"

"Well, you've both been away so long - " said Iska.

"And with all the changes on the estate. Life here is going to be very different in future, with the public swarming all over the place," said Den, "Maybe he has reservations about it all."

"Maybe he has. Don't we all? But so what? We've not exactly had a quiet life all these millennia, have we? Monrosso is the most civilised existence I've ever had," said Iska. "Maybe it's the assassinations. Cold-blooded murder is very different from war."

"And so is being tortured to death – repeatedly," said Lucrezia. "I've tried and tried to convince him it's not his duty to rid the world of terrorists single handed - "

"We should try to talk to him," said Kerallyn. "Find out what's bugging him."

"Why don't we do what he asks and put him out of his misery," said Justin. "That leg must hurt like Hell. We can't just leave him lying there - "

"Well, we're not doing him much good in here. Lets go."

"Who's there?" asked Cesare, through gritted teeth.

"For pity's sake open your eyes, you mugwump," growled Iska. "We know there's nothing wrong with them. We all had a good look while you were out cold with Krisha's brandy cocktail. Stop pretending. You're giving us all the creeps."

"Please, Chez," said Lucrezia. "I can't bear to see you crashing around like a great blind bear. It breaks my heart. Stop this pantomime and be your old self again."

"Just give me the coup de grace and you'll never have to look at me again."

"What! What do you mean? What are you trying to tell me? You've had enough of me? You're going to disappear: start a new life on the other side of the world? Why? What have I done wrong?"

"You've done nothing wrong. You've been the best partner a man could ever wish for. You've been amazing. I'm just not worthy of you."

"Cesare, you've just saved Monrosso."

"Krish, it was you who saved Monrosso. Why did you think you could hide that from me? You don't need help from me. Remember the old days, when we ruled empires? I'd come back from a war to find the empire much stronger and more peaceful than I'd left it, but you refused to take the credit. You'd hand me back the crown without the slightest hesitation. I'm not worthy of you, and I'm worn out, Krish, worn out, burnt out. Time you had a fine new champion."

"Cesare, you're just exhausted. You've taken on far more than any one man ever should. How could I find another man who's half as good as you? But I don't care how good or bad you are. You're the man I love. I've loved you faithfully for twelve thousand years. Stay home and rest. Let me show you how wonderful life could be. Or are you sick of me? Tell me the truth. I'll go away, out of your sight, if that will make you happy."

Her voice broke and she raised a hand to catch her tears. "But if ever you need me I'll be there. Just whistle."

"So, are you sick of all of us? Where are you planning to go? Where will you find another place that's half as good as Monrosso? Why deprive yourself? It belongs to you. If you're sick of us we'll move out," said Iska.

"Don't do that," said Cesare. "I'm relying on you both to take my place, be Krisha's new champions. I know how you love her, admire her. You'll be worthy of her. She'll be in far better hands than mine. I'm going soon. I don't need the coup de grace. Goodbye. Thank you all for the joy of your company. Live long, prosper and be happy."

"I don't understand. Where are you going? Tell me where you're going. I need to know you'll be alright. I can't sleep if I think you may be in trouble."

"I'm going where I'll never have troubles again. I'm going way out into the Great Blue Yonder. I've ordered every one of my atoms to find a star all of its own. I'm going to spread myself out across the whole of the Milky Way. My atoms will never be able to pull together again. Goodbye. Thank you for everything. Goodbye."

The Immortals stared at each other in horrified disbelief.

"Kerallyn, is that possible. Can he really do that?" asked Justin.

"Oh, Great Hadron Collider!" she gasped. "I can't think of any reason why not. All that research I'm doing – it's a complete waste of time. He knew all along how to do away with himself."

"Oh, yes," said Den. "He told me he'd been tempted to call it quits aeons ago, but he'd sworn to be Krisha's faithful champion forever, and he couldn't bear the thought of leaving her alone and defenceless. Now he thinks we can take his place - "

"Den, for Goodness sake, I'm not going to try to hold you to that. You have your own life to live."

"I would count it the greatest possible honour," said Den, "but are we going to let this great big idiot walk out on us like this? We're going to talk some sense into him, aren't we? Listen, you wrong-headed dope -" He leaned over Cesare and shouted in his face, then turned to stare at them. "He's died on us. Now what are we going to do?"

There was a very long silence.

"Are you sure?" asked Lucrezia.

"I can recognise a dead man when I see one," said Den.

"Krish, sit down, lovey," said Kerallyn. "You look like death. Justin, why don't you make some tea. I think we all need it."

"Let's see if anything down here is operational," said Den. "The workmen may have fixed up a coffee maker. Oh look, a kettle!"

"There are two Englishmen on the team," said Lucrezia dully. "How can we talk him out of it if he's dead? What are we going to do? We've got to save him somehow. It's so good living here; surely we could make him happy If he stops doing hideous jobs for the Americans."

"Now we have all this money there's no need to go bounty hunting. He knew that perfectly well, though, didn't he?" asked Kerallyn.

"Maybe he was too disturbed to take it in. But we're wasting time. I've got to stop him going off into the Great Blue Yonder. How can I do that? Will any of you help me? Or are you content to let him go?"

"Let him go? Not on your Nellie. He can't opt out like this. He's our king, our champion. We've got to drag the crazy beggar back," growled Iska.

"Yes, he can't go on his own." said Den. "We were all together when we first became immortal. When it's time to go we should all go together, all thirty of us."

"That's a nice idea," said Lucrezia, "and surely it's irresponsible to go without winding up our affairs. What would happen to our people, and the animals? He took on that last assassination for the sake of everyone here at Monrosso. Why does he want to leave us now?"

"We could debate that till the cows come home," said Kerallyn, "but do we have the time?"

"We've got to concentrate on getting him back, but how? There are billions of stars in the Milky Way. How can we possibly get him back from there?" said Den. "Has anybody any ideas?"

CHAPTER 36

A DEADLY DILEMMA

"What do we know about dying?" asked Lucrezia, in her solemn boardroom tone.

"Do you mean humans or us?" asked Justin.

"Us, of course." Den gave him a withering look. "Actually, come to think of it, precious little. Well, you can't watch yourself die, can you? I've thought of trying to rig up a camera, but, well - "

"Iska? Kerall? No?"

"I'd never been close to another Immortal before I met all of you," said Kerallyn, "but you and Chez must have seen each other die, surely."

"You'd think so, wouldn't you, in all these thousands of years. I've watched him resurrect hundreds of times, sat beside him, willing him to wake up fit and well, but I honestly can't remember actually watching him die. Isn't that odd. He normally died miles away on a battle field."

Justin took a deep breath as realisation dawned. At last he had something important to contribute, something vital he knew that these all-knowing Immortals didn't.

"I watched you die, Lucrezia, and then I watched you both resurrect." It was good to watch their faces, so eager, so expectant.

"Tell us! Tell us!" He had their wholehearted attention.

"Well, first your body smoked away. It was beautiful golden smoke that sparkled like frost. It flowed out of my flat under the entrance doors. Then, next morning, I found you both lying on your ruined four poster bed, glowing like fires. When the bells rang for eight o'clock you stopped glowing and woke up, perfectly normal."

"Smoked away?" said Den. "Well, that explains a lot. I've always wondered how my body gets from A To B. I often wake up miles from where I died. Couldn't make sense of it."

"So, we need to find a way to stop Chez from smoking away. Is he smoking now?" asked Lucrezia.

They pulled chairs up to his table and sat staring raptly at his body.

"How long was it before I began to smoke?" asked Lucrezia.

Justin reran the scene anxiously. "Well, I wasn't there all the time. Chez dragged me off to the B House to clear the rubble off your bed so you didn't materialise under the rubble again. You were alive when we left but when I got back you were smoking like crazy."

"Think, think," said Kerall. "How long do we have before he smokes away up into the Blue Yonder? If he manages to fly up there I can't see how on earth we could ever get him back."

"I think we have to assume we have hardly any time at all," said Lucrezia. "So, let's forget the Blue Yonder and concentrate on stopping his smoke from getting away."

"Build him a coffin," said Iska. "We could use these table tops. Quick, let's put him in that corner and wall him in."

The women stood back while the three men pulled the body off the table, dragged him unceremoniously across the floor, and pushed him up against the wall. Then they boxed him in with table tops.

"Looks like a Roman tortoise," said Justin.

"A what?" asked Den. "Oh never mind. Now what?"

"That's never going to stop him smoking away, is it?" said Lucrezia. "Look at all those gaps. It will leak like a sieve."

"We could block the gaps up somehow," said Justin.

"The smoke may be as fine as atoms and molecules," said Kerallyn. "He said he'd programmed every one of his atoms separately, didn't he? They might whizz through wood as if it wasn't there. It has lots of pores."

"So," said Den, "how do we capture atoms? You need extremely powerful magnets or lasers to keep the atoms inside a particle collider. It costs billions and takes years to build one of those things."

"He's not going to be able to power up his atoms like a particle collider, is he?" asked Lucrezia.

"Did Krish explode?" asked Den. "Did her smoke move very fast?"

"Explode? No, nothing like that. Her smoke just sort of ambled out along the corridor: didn't rush like a whirlwind or anything."

"I've only studied radio-active atoms," said Den. "Class A radio-active atoms just floating about can be stopped by a sheet of paper. We could try wrapping him in paper, but I somehow doubt if that would stop him. Class B radio-actives can be blocked by a sheet of plastic. He couldn't be more radio-active than that, could he?"

"Human bodies aren't very radio-active, are they? They don't trigger Geiger counters, unless they've had radio-active markers put inside them," said Kerall.

Iska pulled out his phone. "Reception's pretty poor in here. Come on, phone, tune in. Got to get through to the Dairy."

"You're not fussing about your blasted animals at a time like this," fumed Den.

"No, idiot, I'm fussing about plastic sheeting."

Justin quickly crossed over to the old radio phone on the counter. It was still live. Nobody had thought to switch the system off now the rest of the House had had its landlines restored. "Hello, Dairy, hello! Mr Romanov wants to speak to you. Iska! Iska! Come over here. I've got your staff on the line."

"We could make him a lead coffin," said Den, "Make sure we'd pinned down every kind of radiation. Plenty of lead on the roof. We don't want to risk something vital getting away and leaving us with some kind of zombie."

"When the body swells up the coffin will come apart at the seams," said Justin.

"Swells up? You didn't say anything about swelling up," said Den accusingly.

"Didn't I? Well, you certainly did swell up, Lucrezia. You were absolutely huge. Gave me an awful fright. Then, when I found the two of you resurrecting you shrank and shrank before you woke up."

"Well, that seems to rule out coffins," said Lucrezia with a shudder. "We need something with lots of room to expand. The only thing I can think of is a huge plastic bag, but where on earth could we find enough plastic? He's nearly two metres tall."

"Right," said Iska, striding back towards them. "Got the plastic sheeting. It's on its way. There's lots left on the roll. We use it for binding the hay bales."

"Is plastic good enough?" asked Lucrezia. "Iska, when you butcher animals and get the meat ready for the shops, what do you package it in?"

"Plastic bags, plastic trays and plastic sheets."

"Do they leak blood or anything?

"Leak? No, of course they don't leak. Plastic is good for any kind of flesh or liquid. If it has a good seal it never leaks."

"Air can't leak out either," said Justin. "Balloons only go flat when the seal isn't airtight. Surely nothing can be finer than thin air. Plastic sounds the best idea."

"Did you tell your people to get a move on?" asked Den.

"No, I told them to faff around all day. They're going to rush the first sheet up so we can settle him on it while they cut another three or four. Will that do for you, Your Majesty?"

"Full marks, Iska, sweetheart," said Lucrezia. "How wide is it?

"A metre and a half. We'll need to join a few sheets."

"Where shall we put him?" asked Den, "Back on the table?"

"No, lets have a clean one," said Lucrezia. "That one's covered in blood. And that's another problem: he's lost an awful lot of blood. We don't want him back all shrivelled and dessicated. We need to find as much of his blood as possible. We can sponge up all the blood in here and put the sponges in with him, hope the blood turns to smoke as well and finds its way back to him."

"We should try to find that big lump of flesh he's lost as well. It must be stuck to the ploughshare," said Iska.

"Hope nobody's washed it away," said Justin.

"No water supply nearby," said Iska, "But something might try to make a meal of it. Better see to that fast."

"We'll organise some bloodsucking," said Den. "I've got a wet vac, battery powered. Who could we use? We all need to stay here."

"Luigi might still be upstairs. He's delivered calves and kids, so a bit of blood shouldn't squeam him. He can take a bucket for the pound of flesh. I'll go look for him."

"I'll go find the vac." said Den.

"We need something to join the plastic sheets together, make a bag," said Kerallyn. "An iron should do the job, melt the plastic and fuse the edges together. I'll go and ask Mrs Lepanto for one."

Suddenly the cafeteria was almost empty. Lucrezia walked over to the tortoise and stood with a hand to her face. Justin pulled her head onto his shoulder and stroked her lovely hair.

"I'd be your faithful champion any day," he murmured. "I'd always come running when you whistled. I know, I know. Cesare's the only man for you. We'll save him for you somehow, talk some sense into him, never fear."

"You're very sweet," she whispered.

CHAPTER 37

THE LONG NIGHT

"What time do you normally resurrect?" asked Kerall.

"Eight o'clock in the morning," said Lucrezia. She sighed. It was only seven in the evening. "We could have a thirteen hour wait. I suppose we should be glad that things are not going any quicker. At least it's giving us time to sort ourselves out, get our act together."

Kerall walked slowly around the corpse again, carefully checking all the joins in the plastic sheeting. She took the peg out of the five centimetre blow hole and squeezed out a little pocket of air.

"Anything happening?" asked Justin.

Kerall shook her head. "He looks so peaceful, as if he's just asleep. No man should be allowed to look so gorgeous."

"We can't all be beautiful," said Justin, ruefully.

"You're gorgeous too," she smiled, "In your own way. Give me a hug."

"Food!" called Den, kicking open the door. "Come help me bring it in. No, you stay there, Maria. Just pass me the dishes. Smells delicious. Thank you very much."

"Please, Mr Fermi, can I see the Marquis? He's getting better, isn't he?"

"Judy, I've told you three times that only the family are allowed down here at present. If I have to tell you again I'll have to ask you to go home. I'm sorry, Mr Fermi."

"Judy, you did a very good job running up here to get help for the Marquis. The family are grateful to you, but he needs peace and quiet, so absolutely no visitors until further notice. Back upstairs, please," said Den.

"It might be wise to help that girl to find another job," said Kerall, "for her own sake as well. She's not going to simmer down while she can see him every day, is she? She's so infatuated she might do something crazy. No, don't ask me what. I'm a psychiatrist, not a clairvoyant."

"Oh, Moussaka," said Den, as they unloaded dishes onto a table. "Very nice. Pity you can't join us, Chez."

"We could save him some for breakfast," said Iska. "He's such a greedy guts he'll eat it cold."

What did all that signify? Justin wondered. Did they believe he wasn't really dead, or were they certain they could bring him back somehow? What now? Somebody was banging on the garden door.

"I'll go," said Iska. "Thanks, Luigi, that's brilliant. Better than I expected. Rooks? Yes, I was worried about them. Good for you. Well done. Look, It's way past your supper time. Do you like Moussaka? Can you stomach a plateful? You'll probably feel better with some nice easy food inside you."

Den strode across to the door. "Here you are, Cook's special Moussaka. Put hair on anybody's chest. Go eat it in the fresh air. Leave the plate outside the door, and thanks again for all your efforts today. Goodnight." He closed the door. "Nice lad you've got there, Iska."

"What did he want?" asked Lucrezia.

"Have you finished eating?" asked Iska.

She nodded."Why?"

"He's brought back the wet vac and the bucket. The rooks had spotted a good meal but our intrepid cowboy drove them off and rescued most of it. He looks a bit wan, but he's a tough lad. I bet he'll keep the Moussaka down alright."

"I'll put it all into the bag," said Kerall. "No, Krish, it's easier for me. Save me some pudding. Look, I may be only a humble psychiatrist but I've handled plenty of blood and guts over the years." She went over to the door to pick up the bucket, then took it over to the corpse. She soon went over to the counter and began rooting about.

"What do you need?" asked Den.

"Scissors. Can't get it in through the blow hole."

"Make the hole bigger," said Iska.

"No! No! Once the fun starts we could have trouble keeping him in." said Justin. "Don't make it harder."

Den picked up the phone. "Hello, kitchen, Heidi? Yes, we need some scissors down here below stairs, and a big funnel, one with a nice wide spout. Can you bring them down? No, you yourself, not - I repeat - not Judy."

"What are they for?" asked Lucrezia, as Den collected them at the door and took them to Kerallyn. "I had some scissors - "

"Don't ask," said Iska. "More pudding?"

It was lucky most of them had already eaten their fill before this flesh and blood arrived, thought Justin, watching Kerallyn at work. Wonder how much flesh the rooks had eaten before poor Luigi drove them off and pulled it off the ploughshare. She had finished pushing strips of it through the gap and was now inserting the spout of the funnel. He swallowed hard and removed the transparent bin from the wet vac. The blood was black and treacly now. Would that matter, he wondered.

"What about all that stuff?" he asked, as Kerall scooped out grass that had already blocked the funnel. "Shall I get a rubbish bin?"

"No, No, I'll put it in the bucket for the moment, then push it in here later when all the blood is in." She poured the lumpy blood very slowly through the funnel, careful not to spill a drop.

"Will the beetles matter?" He could see at least ten

assorted beasties exploring their new home.

"Dunno," said Kerall. "Shouldn't think so. We must all have died and resurrected with beetles and lots of other passengers over the millennia, mustn't we? Maybe some of them ended up catching immortality as you did. I suppose we'll never know."

"Why push in all this grass?"

"It's soaked with blood. He's lost an awful lot of blood, so every drop is precious. As he resurrects he'll draw it back into himself. I've bled to death now and then, soaked the carpet, but next morning the carpet was clean as a whistle. I'd soaked it all back up."

"There must be a lot of blood still soaked in the soil by the ploughshare. Can he pull it back from so far away?"

"Don't know. Maybe someone should go look, see if Luigi missed anything." Kerall cleaned the vac bin, the bucket, the funnel and her hands very carefully, then pushed the cloth into the bag. "Every drop counts."

"Justin, come tell us again very slowly what you saw when we died," called Lucrezia. "This might be our only chance to save him. We've got to be as well prepared as possible.

"Well," said Justin, "it was about nine-thirty pm when Chez dragged me off to move the rubble in the Borgia House. You were alive then, but he must have known you were dying and he didn't want you materialising under the rubble again. I left him there alive about two-thirty in the morning. When I got home about two forty-five you were unrecognisable, a huge swollen thing, glowing like a fire and smoking like crazy. I thought I'd gone mad and shut myself in my room. At seven next morning there was no sign of you. I went to the B House to see what had happened to Cesare and there the two of you were, glowing like fires and a bit swollen. By eight am you looked perfectly normal and woke up."

"So, Chez was alive until after two-thirty am and resurrected at eight. He managed the whole cycle in about five hours. Maybe nothing will happen until after two-thirty. We should have time to think of any extra precautions we could take. Think hard. It's now or never."

"What if the bag bursts? He's a big man. I hate to think what he might be like all swollen and glowing like a furnace," said Justin. "You looked scary enough."

"Have we enough plastic left to make another bag?" asked Kerall.

"Three sheets, that's four and a half metres wide. A bit tight?"

"We could make a flat sheet, and nail it to the tables if we have to. I'll plug the iron in again," said Kerall.

"How hot was Krish?" asked Den. "What if he melts the bag?"

"A hose, quick, find some hoses,   Find some taps – and some tap fittings."

There was a mad scramble into every corner of the cafeteria. Justin tore up the stairs and out into the kitchen courtyard. Thank Heavens the hoses were still where he had left them a few weeks before! He unfastened the fittings and took them and the hoses downstairs.

"Anyone know how plastic sheet behaves with flaming heat one side and cold water the other?" asked Den.

Big sighs all round.

"Have any of you resurrected and found your bed all singed, or very hot?" asked Lucrezia.

After a few moments' thought they shook their heads.

"So," said Den, "it's mainly bright light that we give off, not heat. Anyway, we've got a water supply ready if we need it. Anything else? Come on, think."

They thought and thought, to no avail.

"Well," said Lucrezia, "it's a long time till two-thirty. Shall we draw up a duty rota, a pair at a time?

"I'll go up to the attics and find us a blow-up bed or two. These hard benches should keep the duty guards awake," said Den.

"And a few quilts, please," called Lucrezia. "It may get cold overnight. Now, If anyone thinks of anything else they must wake the rest of us up immediately. Right?"

"If we can be off duty first we'll go down and dig up some soil by the ploughshare, Should be some blood in that," said Justin. And there certainly was. All those stories about the ancient gods making humans out of soil mixed together with gods' blood suddenly made sense. They each brought a bucketful of the sticky blood pudding back to the basement.

Cesare's giant plastic bag now looked like a terrarium, a playground for creepy crawlies, with soil to crawl into and grass and leaves to munch. In the middle of it all lay Cesare, beautiful as ever, with a look of perfect peace on his face.

"Woops, we'd better get you out, Catty," said Kerall. She carefully inserted a slender arm and winkled out the caterpillar from amongst the blood and guts. "Some of these little critters can eat through a super-market plastic bag within an hour."

What messy life-forms we humans are! thought Justin, as he settled down on the squeaky little blow-up bed. You can't make anything useful out of a dead human, can you, not a nice table or something. Once dead we're just a big lump of rotting rubbish. Yuk! If you cut a big tree down every bit of it looks nice and neat and has a gorgeous smell, and you can make it into so many beautiful things: sideboards - wardrobes - violins -

289

CHAPTER 38

LAST CHANCE

"Twelve o'clock. Wake up, Justin, Kerall, wake up!"

Justin rolled off the bed onto his knees on the cold stone floor, groaning. "Anything happening yet.?"

"Yes, come look," said Lucrezia. "Just started."

"You should take a rest, Lovey," said Kerallyn. "Go lie on my bed."

"I've the rest of eternity to take as much rest as I like," she said grimly. "This could be my last chance to look at Cesare. "Oh look! Look!"

Cesare's beautiful face was moving, as if seen through a distorting mirror. Justin's stomach lurched as the eyes drifted apart and the mouth grew wider and wider. His head was expanding unevenly. Wisps of smoke were drifting from his nose, his ears. Soon the face was hidden by a whirl of smoke. Smoke was pouring from his clothes and spreading around the plastic bag,

Lucrezia swallowed hard and turned away. "Iska, Den, get some rest for half an hour; then it might be all hands to the pumps."

"Here, sit down before you fall down," insisted Kerallyn, dragging a chair up behind Lucrezia. "This may take hours and hours. We need rest breaks or we might all fall asleep at once, then, well, who knows."

It was a very long, slow process. By one-thirty it was no longer possible to recognise the shape of a body: it was just a swirling soup, churning as if stirred by a huge invisible paddle. The soft golden glow was now a glaring orange-red.

"We need sunglasses," said Justin. "It's going to get a lot brighter than this."

"I'll go up and get some for the three of us. Den, Iska, you'd better go get yours," said Kerall."

By two-thirty the contents of the bag were so bright even the sunglasses weren't enough. They had to squint through slightly parted fingers.

"It's not too hot to touch," said Den. "Keep a hand on the bag to check it's not under too much strain. "

"Feels near bursting at the seams to me," said Iska.

"You could be right," said Den. "Emergency sheet, quick!" He leapt across the room and grabbed the flat plastic sheet. Five pairs of hands grabbed it and hauled it into place. Kerall grabbed hammers and passed them around. Lucrezia scooped up handfuls of nails and scattered them within reach of everyone.

So we nail the man down, and we nail the man down, Justin sang silently. This had to be a nightmare – and a pretty weird nightmare. He banged nail after nail into the edge of the wooden table.

The edge of the sheet was pulling so hard it tore itself away from the nail. Desperately Justin tried to halt it with another nail further in, and another and another. They'd bitten off more than they could chew.

"Stoppit, Cesare, stoppit! We're not going to let you go!" screamed Lucrezia. "We need you. Don't leave us. We need you, we love you, don't leave us!"

Bedlam ensued. Everybody screaming, banging nail after nail into the table as the huge plastic balloon stretched ominously.

Suddenly everyone held their breath. The roiling stew within the balloon had slowly ground to a halt. They stared at each other, hope dawning. Had he given up trying to fly away?

Oh, no!" wailed Kerall. "He's found a way out. Look!"

They followed her gaze around the room. Every corner was lit by sparkling golden smoke. It was all for nothing. They had failed.

"The doors!" shouted Justin. "Stop him getting out under the doors." He grabbed a quilt, dragged it to the out door and began to wedge it into the gap at the bottom. Then he stopped, bewildered. The smoke was not going out: it was coming in. "Look, come look! This smoke is coming in. What's going on?"

"It must be his blood coming home," said Lucrezia. "All the blood he lost that's still out there. It must be coming home to him. I've seen that happen before. He must have stopped trying to fly away: he's begun to resurrect, right here. Come watch: I think it will all head straight for him."

The contents of the bag were invisible now, hidden by a layer of dust like a fall of sparkling golden snow.

"It can't get in," said Kerallyn. "He might wake up very sick, with so much blood missing. That's what he died of, isn't it? Quick, cut a hole in the bag."

"Do we dare?" asked Justin. "He could be trying to trick us, come flying out like a hurricane if we give him the chance."

"He wouldn't do such a sneaky thing to me," said Lucrezia. "Those in favour of cutting the bag open?"

Slowly, tentatively, all hands rose. Iska took a deep breath and opened his Swiss army knife.

They held their breath and watched, as the golden snow began to creep slowly and steadily towards the hole and into the bag.

"Welcome home, Cesare," breathed Den. "Cut a hole for his face. We don't want him to suffocate as soon as he wakes up."

CHAPTER 39

THE MORNING AFTER THE NIGHT BEFORE

It was half past seven. Bleary eyed but wide awake, they warmed their hands on their mugs of tea.

"Leave enough milk for Cesare," cautioned Lucrezia.

There was a tensing of lips on every face.

"How should we play this?" she asked. "He might be furious with us. He sounded very keen to go off to the Great Blue Yonder. Do you think I should keep out of his sight? He must have had his fill of me. I've been nagging and bullying him for thousands of years. It's amazing he's put up with me so long. If he thinks I've gone maybe he'll feel less desperate to leave."

"None of us have been very sympathetic since he lost his sight, have we?" said Kerallyn.

"Well, we knew he was making it up. It was just a pantomime. He was making fools of us," protested Den.

"So, you don't believe in hysterical blindness, then?"

"Cesare, hysterical? He's as tough as old boots."

"That's nonsense. He's just the bravest, the kindest, the most uncomplaining man you'll ever meet," cried Lucrezia. "And we've driven him to suicide."

"Maybe we should clear off, then" said Iska. "Get out of his hair. Tell him we're very grateful for his hospitality all these years. Tell him we're on call, back like a shot if he needs us."

"Maybe Iska's right," said Den. "Tell Chez we'll miss him. We'll always come running if he needs us. Let's go have breakfast and then pack our things."

Lucrezia watched them go, then hovered by the door.

"Somebody should stay, make sure that he's all right.

He shouldn't have to wake up all alone."

Justin sighed. "I'll stay. He's no axe to grind with me. I'll tell him he has a clear field, a fresh start. That might soften the blow that he's still here, in this rotten old world, that he hasn't managed to get away,"

"You will call if he needs us, if you need us. I'll be straining my ears." Reluctantly she turned and trudged slowly up the stairs.

"He's mistaken me for Krisha more than once. If he's annoyed with Krish and he sees me he might - well, it might not be - exactly helpful." said Kerallyn lamely.

"Go on," said Justin resignedly. "Save me some breakfast if you can."

"Will do. Good luck."

"I guess I'll need it," breathed Justin. How on earth did I get myself lumbered with this?

The lights went out. Bright sunshine through the windows lit the first two metres of the big half-built cafeteria, but the inner part looked gloomy. They were definitely going to need the new lights that were being installed. But they weren't working yet. How could they have gone out. Realisation dawned. Cesare was no longer glowing. The sparkling golden light had gone. Cesare was just a body lying on a table. Heart sinking, he cautiously moved closer. The body turned its head almost imperceptibly and blue eyes interrogated him.

Justin cleared his throat. "Morning, Cesare. Your eyes seem to be working now. Good show."

"So they are. That's useful." He yanked the plastic sheeting from his head and torso and sat up, staring at the wrinkled bag encasing him. "Fascinating. Who's idea was this?"

"Joint effort. Big brainstorming session. You'd have

been amused. One idea was to pull the lead off the roof and wrap you up in that. Gift-wrapping you in fancy paper was another idea that fell by the way-side. And so did walling you up in that corner with table tops. Looked like a few Roman soldiers in a defensive tortoise."

Cesare laughed, wriggled his legs out of the bag and swung them off the table. No sign of injury, thank Goodness.

"Phew, it was hot as a sauna in there. Huh, somebody's wasted a lot of nails and ruined a good table. What was the idea of all this?"

"It was all your fault," said Justin. "You grew as big as a Tyrannosaurus Rex and fought like a fiend to get out of the bag, so we nailed the man down and we nailed the man down. Hey ho!" Euphoria engulfed him. It was all over. They had won, hands down. Cesare was his old self again. All was right with the world.

"Why on earth did you want to bag me up?"

"The others loved the idea of going off to the Great Blue Yonder. They want to come as well, but they think you all ought to stay and settle your affairs properly before you go. Only fair to the people and the animals - especially the wolves - and the alpacas - and the highland cattle and - " He tried to keep his face straight, but Cesare didn't. He snorted with laughter.

"Don't forget the horses and the goats, especially Rumpelstiltskin. Looks as if you've had a party down here. Anything left? I'm starving."

"We saved you some Moussaka. Shall I get Cook to warm it up?"

"Don't bother. I'll eat it cold." He reached out for the plate.

"Yes, that's what Iska said you'd do."

"Well, so would he. I'm not the only greedy pig around here. Where is everybody?"

Justin ran a few possibilities frantically through his head and rejected the lot of them. Let fate take its course, he thought, and time for another little lie.

"Gone up ahead to tell Cook you're dying for a good fry-up. The Full English should be ready by now. Hey, don't you think you ought to put some clothes on? Don't give Cook a fright." Only a man who knows he looks stunning could be so unselfconscious in the buff, thought Justin enviously, as he helped Cesare root about in the plastic sheeting for his clothes. He shook the T-shirt, sending grass, soil and beetles flying.

Cesare pulled on his jeans and looked at the tattered leg, cut open from the ankle. He sat on the table and swung his leg up onto the surface. Fascinated, Justin watched him carefully manoeuvre the two sides of the ragged fabric together, then blow on them. The threads moved, actually moved together, apparently of their own volition. He inspected the mend, and smoothed it with his fingers until every broken thread had disappeared.

"That's better," he said, "but tricks like that should only be done among friends, Immortal friends. Now, do I meet with your approval? Yes? Let's go." He stormed through the door and took the steps three at a time.

Justin followed, weak at the knees. Would the others catch the happy mood, behave as if nothing distressing had happened? Well, they were far from stupid, and some of them could read each others' minds. He put on a big grin, walked into the kitchen and shouted, "Voilà! Look what I found lazing about below stairs. Hope you saved him some breakfast! I told him you were starving hungry and might have scoffed the lot. And you were right, Iska: he did eat that Moussaka cold."

"Greedy pig!" laughed Iska. "What did I tell you?"

"Takes one greedy pig to recognise another," laughed Cesare.

BOOK 3

GOODNESS KNOWS WHERE!

Justin Chase's computer-hacking skills prove useful once again, creating a mutual help society by enticing out survivors lying low in interesting places. Learning to skipper a freighter in some of the world's most dangerous waters is an unexpected challenge he must face.

Meanwhile, the survivors at Monrosso find thrilling new ways to make their idyllic country estate pay its way, despite the protected wolves, and rebuild their palace in Rome as a fortress for survivors, disguised as a tourist attraction.

But first, superhuman methods are needed to rescue poor Kerallyn from her perspex coffin in Rome's Modern Art Museum – without giving the game away.

BOOKS 4, 5 AND 6

Books 4, 5 and 6 bring in more survivors unearthed by Justin Chase, not all of them welcome additions to the family. And how do you get rid of a survivor if he proves to be a threat to all mankind?